LOVE DRUGS

LOVE DRUGS

The Chemical Future of Relationships

BRIAN D. EARP &
JULIAN SAVULESCU

Redwood Press
Stanford, California

STANFORD UNIVERSITY PRESS
Stanford, California

Printed in the United States of America on acid-free, archival-quality paper

Library of Congress Cataloging-in-Publication Data

Names: Earp, Brian D., 1985- author. | Savulescu, Julian, author.
Title: Love drugs : the chemical future of relationships / Brian D. Earp and Julian Savulescu.
Description: Stanford, California : Redwood Press, 2020. | Includes bibliographical references and index.
Identifiers: LCCN 2019009056 | ISBN 9780804798198 (cloth ; alk. paper) | ISBN 9781503611047 (epub)
Subjects: LCSH: Psychotropic drugs. | Love. | Couples—Psychology. | Interpersonal relations. | Medical ethics.
Classification: LCC RC483 .E23 2020 | DDC 616.86—dc23
LC record available at https://lccn.loc.gov/2019009056

Cover design: Michel Vrana

Text design: Kevin Barrett Kane

Typeset at Stanford University Press in 11/15 ITC Galliard Pro

To Mom and Dad, with all my love.

—B.D.E.

To Miriam, who has made everything possible.

—J.S.

To love somebody is not just a strong feeling—
it is a decision, it is a judgment, it is a promise. If
love were only a feeling, there would be no basis
for the promise to love each other forever.

—ERICH FROMM

CONTENTS

LOVE DRUGS

CHAPTER I

REVOLUTION

"OXFORD ETHICISTS PROMOTE MDMA to combat divorce." At least, that's what a blogger at *Dose Nation* said we were doing when we first started writing about the chemical enhancement of love and relationships. The blogger was referring to an interview we'd done with *The Atlantic*, where we argued that certain psychoactive substances, including MDMA—the key ingredient in the party drug Ecstasy—might help some couples improve their connection if used in the right way. The truth is, we were not promoting the use of MDMA outright. We were calling for research into this possibility while exploring its ethical implications for individuals and society. The same aim, bolstered by the latest data and insights from the cutting edge of bioethics, applies to the book in your hands.

This time our call for research is more urgent. MDMA, along with psychedelic drugs like psilocybin (from "magic" mushrooms) and even lysergic acid diethylamide (LSD), are moving quickly into the center of mainstream medicine. Receptive pieces by hard-nosed journalists and science writers are coming out almost daily. In the *New York Times* alone, there have been articles on the "Promise of Ecstasy for PTSD," the benefits of "Magic Mushrooms" for depression in cancer patients, and the potential of LSD as a treatment for alcoholism and anxiety. Ten years ago such coverage would have been confined to fringe sources. But now that phase 3 clinical trials for some of these drugs are in the pipeline (or already started), with a stream of smaller studies practically flowing out of labs at Imperial College London, Johns Hopkins, and other major institutions, things are starting to change. Phrases like "paradigm shift" are beginning to creep into the titles of research reviews on the science of drug-assisted psychotherapy.

Our *Atlantic* interview came out in 2013. This was two years after the first pilot study on MDMA for post-traumatic stress disorder (PTSD) made its way to publication. PTSD is an often disabling condition that can develop in response to violence, including exposure to killing and bloodshed in warzones. It has been a widespread problem in the military for decades, reaching "epidemic-level heights" with the wars in Iraq and Afghanistan. As the *Washington Post* reports, between 11 and 20 percent of soldiers in these wars suffer from PTSD, which has "ravaged lives and broken up marriages." Trauma affects more than individual survivors. It affects friends, family, and romantic partners, including spouses. Some spouses can hardly recognize the person they married, a once-familiar life partner now prone to unpredictable panic attacks, anger, and even physical aggression.

This raises a simple but important point. Successful treatment of PTSD, by whatever means, can have positive implications for relationships. But this shouldn't be a side effect or afterthought. Medical research needs to take interpersonal factors, the space *between* people, into account—from the initial design of studies, to the collection of data, all the way through to the write-up of the final report. In this book, we focus on romantic relationships, but the point holds more generally. Introducing a drug or other medication into the life of an individual also introduces it into the lives of those who love and care for them. Under current treatment norms, the effect on others happens indirectly. But what if it happened directly, with intention? What if the goal was to improve people's lives along a relational axis? Romantic partnerships might then become a worthwhile focal point when considering certain drug-based interventions.

Of course, relationships themselves can be traumatic. Sometimes this means they should end. But in other cases there may be significant value in trying to restore a broken connection. Especially when the distress is rooted in outside factors, as is often the case with PTSD, romantic partners may be desperate to find a solution that allows them to maintain the relationship. Yet with PTSD, unprocessed traumas can be difficult to put into words, much less talk about productively with a partner. This is part of the reason why traditional talk therapy, and even conventional drugs like Zoloft and Paxil, have done so little to stem the tide of symptoms associated with this condition. "If you're a combat veteran with multiple tours of duty," says John Krystal, a PTSD specialist and the chair of psychiatry at Yale, "the chance of a good response to these drugs is 1 in 3, maybe lower. That's why there's so much frustration and interest in finding something that works better."

This brings us to MDMA and why it seems to show such promise: it sidesteps language. Current research suggests that MDMA temporarily reduces hair-trigger fear responses by working directly on emotional centers of the brain. Objectively, it causes the release of serotonin and other neurotransmitters and keeps them in play in the clefts between synapses. Subjectively, as one writer put it, "it imbues users with a deep sense of love and acceptance of themselves and others—the perfect conditions for trauma therapy." Findings from the early MDMA studies bore this out. Over two-thirds of participants suffering from chronic, treatment-resistant PTSD no longer met the criteria for the disorder twelve months later. Standard therapies do not come close to this degree of effectiveness.

♥ ♥ ♥

C. J. Hardin is one of the soldiers who has benefited from MDMA-assisted therapy. After three tours in Iraq and Afghanistan, he was numb from the stress and violence. Then his marriage fell apart. Depressed and alone, he retreated to a backwoods cabin in North Carolina. He turned to alcohol to drown his pain. He considered suicide. He tried every available treatment for PTSD, from group therapy to nearly a dozen medications.

"Nothing worked for me, so I put aside the idea that I could get better," he told a reporter. "I just pretty much became a hermit in my cabin and never went out." But MDMA-assisted therapy, he says, "changed my life. It allowed me to see my trauma without fear or hesitation and finally process things and move forward."

C.J. says he saw "a profound difference" in his symptoms after the very first treatment. After three sessions of therapy, "my score on the Clinician-Administered PTSD Scale went from 87 to 7 and I no longer qualified for a diagnosis of PTSD."

"We can sometimes see this kind of remarkable improvement in traditional psychotherapy, but it can take years, if it happens at all," said Dr. Michael C. Mithoefer, the psychiatrist who conducted the early trials, in an interview. "We think it works as a catalyst that speeds the natural healing process."

What about C.J.'s relationship? By the time he got his life back, it was too late for his marriage. There is no way of knowing how things might have gone if he'd found an effective treatment earlier. Maybe his marriage needed to end. Maybe it could have been saved and gone on to blossom. The only way to answer such questions is to ask them in advance.

Dr. Michael Mithoefer is aware of the need for such research. With his wife, nurse Ann Mithoefer, he has started a tentative program looking into MDMA-enhanced psychotherapy for couples where one of the partners has PTSD. As of late 2017, they had worked with three married couples. A tiny sample to start with, but it could be the seed of something bigger to come. "The focus," Dr. Mithoefer says, is on "tackling the affected partner's PTSD and addressing issues it may have created in the relationship."

It is a good place to start. But not all relationship problems stem from full-blown mental illness. Could MDMA-assisted counseling be helpful for a wider range of issues in the not-too-distant future? We tackle that question in detail later in this book. For now, the focus is on treating the most devastating of psychiatric conditions, an approach that is winning over crucial gatekeepers in the government.

In fact, during the same year the Mithoefers started their work with married couples, MDMA was granted "breakthrough" status by the U.S. Food and Drug Administration, one of the nation's top health and safety watchdogs.

The following year, a phase 2 clinical trial was published in *The Lancet Psychiatry*, reporting that MDMA-supported psychotherapy was effective at reducing symptoms of PTSD in a small sample of war veterans and first responders. A much larger, phase 3 trial is now underway. As for safety, the accumulated evidence suggests that short-term, limited use of MDMA—with professional supervision in a comfortable, therapeutic environment—carries a low risk of physical or psychological harm. The reputation of MDMA as a dangerous party drug is melting away with each carefully conducted study.

Early days

When we started our work almost a decade ago, MDMA was still mostly known as a legally forbidden shortcut to dance floor euphoria. And the risk of serious harm in such contexts *is* much greater than in therapeutic settings—up to and including death, as we will see. As an additional hurdle, our proposal that MDMA might one day be used as an aid in couples counseling, including for clients without a psychiatric disorder, was not so obviously within the realm of conventional medicine as a cure for PTSD. So our *Atlantic* interview turned some heads: two ethicists (of all people) from a conservative old British university (of all places) talking publicly about the potential nonmedical benefits of a currently illegal drug. It wasn't what people expected.

But it shouldn't have been so surprising. After all, the question of whether a drug, or anything, should be illegal in the first place *is* an ethical question, and current laws are not always adequately justified. The same goes for the lines that get drawn between "medical" and "nonmedical" substances. Whether we call a drug medicine, or regard it as a form of recreation, or try to harness its effects to improve our lives (what we refer to in this book as enhancement), it makes no difference to the molecules in question. They work on the brain however they work, and produce whatever effects they produce—good, bad, both, or in-between.

Often the legal or medical status of a drug has more to do with quirks of history or politics than with a sound understanding of its actual benefits and risks. Does a drug-induced mood-lift count as medicine? If you suffer from depression and got the drug from your doctor, yes. If you don't suffer from depression and got the drug from your dealer, no. But the boundaries are blurrier than they first appear. In this book, we will explore the ethics of using drugs for *relationship enhancement*, breaking out of the individual-centered, disease-focused model of modern medicine. In short, we will argue that if drug-assisted couples counseling can truly help improve people's relationships, then there should in principle be a way for them to access that experience. And psychiatrists shouldn't have to first make up a raft of "relationship disorders" so that the experience can qualify as medical treatment.

If our claim is justified in principle, what about in practice? A lot depends on the details and on the implications for society. Widespread changes in patterns of drug use or modes of access could have rippling consequences, both positive and negative, many of which would be hard to predict. That is why we need to have this conversation. But whether or not our specific proposal catches on, there are already signs that society is on the fast track to a drug revolution. We think it will start with greater acceptance of mind-altering substances as treatments for recognized psychiatric disorders, like PTSD. It will move from there to the use of such substances to help people in general improve their lives and relationships from whatever psychological baseline they happen to be at—including some that would now be considered healthy or normal.

♥ ♥ ♥

Evidence for this second revolutionary phase is building. One lead comes from the broad success of journalist Michael Pollan's recent meditation on psychedelics, *How to Change Your Mind.* With lessons for "consciousness, dying, addiction, depression, and transcendence" promised in the subtitle—and thoughtfully delivered—the book is not just being touted by aging hippies and panned by everyone else. Instead, it has touched a nonpartisan nerve, as people from every corner of social, professional, and political life are showing a cautious open-mindedness about mind-opening drugs.

Another recent lead is a 2018 study conducted by Roland Griffiths, the respected Johns Hopkins University neuroscientist and drug researcher, which looked at the effects of psilocybin (the active ingredient in magic mushrooms) combined with meditation in healthy volunteers. The results, published in the *Journal of Psychopharmacology*, suggest "enduring positive changes in psychological functioning and in trait measures of prosocial attitudes and behaviors." No disease or disorder here—just ordinary people seeking positive change.

Griffiths and his collaborators believe that under appropriately supportive conditions, psilocybin can "reliably occasion deeply personally meaningful and often spiritually significant experiences." Neither of us is particularly "spiritual," but we have contributed to academic and public debates about the ethics of human enhancement, and we see some overlap between the two ideas. Unless you view the human condition as one big disease state (which seems like a stretch), attempts to improve yourself "spiritually" are not premised on being ill. Whether it's meditation, hiking in the mountains, practicing yoga, or using psychedelics under the right conditions, spiritual practices meet you where you are and aim to get you somewhere better. The same is true of biomedical enhancement: the use of drugs or other technologies to improve so-called normal human traits and abilities.

Seen in this light, our habit of categorizing drugs as either medical or recreational might seem a bit myopic. Some drugs that are used for recreation are risky and addictive, and often make people's lives much worse. Alcohol is a good example. Some drugs that are used for medicine, like prescription opioids, follow the same basic outline. But other drugs that are used for recreation (and now increasingly for medicine), like MDMA and psilocybin, are less risky, generally nonaddictive, and can help make people's lives much better. In our view, when a drug has this potential—to improve people's lives if used in the right way—it shouldn't matter so much whether we call it medicine or something else. The important thing is to learn how it works, to understand its effects, both good and bad, and to clarify the conditions for individuals, relationships, and society under which it brings more benefit than harm.

Something like this view is catching on. People are starting to realize that not all drugs are "bad" and that you don't necessarily need to have a disease or disorder for a chemical substance to help improve your life. In this book, we explore whether drugs can ethically be used to enhance that aspect of our lives we often care about the most: our romantic relationships.

Love, drugs, and marriage

"How LSD Saved One Woman's Marriage." That's another headline from the *New York Times.* The reference is to a self-experiment with the drug colloquially known as acid by novelist Ayelet Waldman, as recounted in her memoir, *A Really Good Day: How Microdosing Made a Mega Difference in My Mood, My Marriage, and My Life.* Microdosing refers to a practice now popular among Silicon Valley types of taking a tenth or so of a regular dose of a psychedelic substance on a somewhat consistent basis (say, every few days). The effects are supposed to be subperceptual: not enough to cause a full-blown trip but enough to feel "sparkly," as one writer put it. And enough, apparently, to bring a troubled relationship back from the brink.

"I was suffering," Waldman writes. "Worse, I was making the people around me suffer. I was in pain, and I was desperate and it suddenly seemed like I had nothing to lose." The way Waldman describes herself and her behavior before her microdosing experiment is not flattering. She would frequently pick fights with her husband for no apparent reason, and then berate herself for having done it. Her frustration would then compel her to lash out again, making her even more despondent—"my shame spiral screwing a hole right through our relationship." Once she started microdosing every third day, she says, she found herself becoming a better listener, calmer and more content, less prone to conflict, more productive, less irritable, more flexible, more affectionate, and more mindful. Not surprisingly, all of this was positive for her marriage.

Waldman had a good experience. But there is reason for skepticism when you are dealing with a single account. Not only can microdosing land you in jail (it's illegal), but there are currently no good scientific data on the benefits and risks of this habit, or even how it differs from a sugar-pill placebo. Anecdotes are not enough. What we need is careful research: certainly, the empirical kind associated with lab coats and clinical trials, but also ethical and sociological studies to make sense of the moral and cultural dimensions of drug-enhanced modes of living and being.

Drug-assisted breakups may also soon be a possibility. One of us (Brian) received an e-mail the other day from a total stranger, written in an Eastern European language. With the help of Google Translate, Brian got into a back-and-forth with his impromptu correspondent, a despondent housewife as it soon became clear. We'll call her Sofia. A heartbreaking picture emerged. Sofia was clearly desperate and seemed to be in a bad situation. She couldn't live with her husband, she said, because he was oppressive and misogynistic. But she couldn't leave him either, because despite all that, she loved him—really loved him—and the thought of splitting up made her despair.

Sofia knew she needed to get out of the relationship, but her heart kept saying no. So she reached out to us for a remedy, some kind of "cure for love," as she put it, that would vanquish her feelings of attachment to her spouse. Freed from the bonds of a love gone bad, she might then try to start her life over with someone else. She was requesting what we call in our work an "anti-love drug."

Suppose Sofia took such a drug. Could it really make her fall out of love? Partly, this depends on how the drug would work—how exactly it would act on her brain, what side effects it might have, and how it would affect her thoughts and feelings. But it also depends on a deeper, philosophical question we will be grappling with throughout this book: namely, what it actually means to *be* in love (much less fall out of it). Some would argue that Sofia couldn't have been experiencing real love, because she was in an oppressive relationship. This is a normative definition of love: it says that the very concept should be reserved for relationships that are essentially positive, good, or healthy. Since love is a highly valued phenomenon, this perspective goes, we should take a moral stand on what sorts of things get to count as love in the first place. If the feelings between individuals in a harmful or abusive relationship are described as love, the worry is that it might legitimize, or provide cover for, the abuse.

It's a compelling argument. We have no problem with people who want to use "love" in this restricted way. But it's a risky argument as well. Once we start defining for other people what love is, even overriding their personal judgments, we can slip into a narrow-minded and paternalistic way of thinking that discounts their lived experiences. This is not just an academic concern. Only a few decades ago (and in many places still today) it was commonly held that love between same-sex partners was a conceptual impossibility, a mistake in thought and language, since *real* love could only occur between a man and a woman. For a depressing illustration of this attitude, look up the YouTube video "Christopher Hitchens vs. Bill Donohue." Then listen to the incredulous groaning and laughter from the audience when Hitchens makes the claim that homosexuality can be a form of love. That debate took place in 2000 in New York City. Not so long ago, and not in some far-off kingdom.

The point is that normative definitions of love often favor the group in power, and their perspective is not always justified—even if they have good intentions. The tendency to "medicalize" love and say that it only really counts if it's "healthy" may be an example of this. An alternative approach, which is broadly the one we take in this book, would be to opt for a more neutral or descriptive route, giving wide berth to individuals to feel and conceive of love in their own way. When we are talking about people's romantic experiences, then, we will mostly avoid couching them in thick theoretical terms or trying to show how they link up to the latest philosophical account of love. That is, we'll often use the word "love" in an informal way and let you fill in the relevant sense according to the context and your own intuitions. And in a similar vein, when we tell you stories about individuals who claim to be in love, we will let them speak for themselves and try to take them at their word.

Some philosophical accounts of love actually support this approach. One of them, defended by the Danish American philosopher Berit Brogaard, says that love is, first and foremost, an emotion: a subjective, conscious, relational feeling that persists through various circumstances and lengths of time which only *you*, the individual, can directly access. In other words, barring special circumstances like obvious delusion, if you sincerely believe you are in love with someone, then you are. One implication of this account, including for the situation with Sofia, is that it *is* possible to truly love someone who is not good to you or who even hurts or abuses you. It's just that this love may be so foolish, harmful, or irrational that you have reason to make it go away—with or without the use of a drug.

Another prominent theory says that love has two dimensions, a *dual nature*. This theory tries to get beneath the emotions we feel when we're in love, and explain where they come from and how they work. The first dimension is biological. It acknowledges that our capacity for love is deeply rooted in our evolutionary history, reflecting basic human drives for sex and attachment. These drives promote our continuation as a species, motivate us to care for vulnerable offspring, and fill a deep survival need for unconditional support. The other dimension is psychosocial and historical. It speaks to the cultural norms, social pressures, and ideological constraints that exist at a given place and time and shape how we think about, experience, and express romantic love in our daily lives.

We will get into this dual-nature theory in the next chapter. We bring it up now to make a point about Sofia. If love, including romantic love, really does have a dual nature, both biological and psychosocial, then it might be possible to *change* love along one or both dimensions. Take the psychosocial aspect first. One way Sofia could try to dissolve her feelings of attachment to her abusive partner would be to intervene in this domain. She could seek support from friends and family, affirm her own self-worth, and deliberately dwell on her husband's evident faults and failings. Talking to an experienced relationship therapist, if she could find one, would also be a good idea.

Indeed, she *should* do those things. If it is right for a relationship to end, then pulling social and psychological levers to create emotional distance and open up a space for healing is indispensable: a drug should never be used as a replacement for these measures. But sometimes these measures will not be enough. Sometimes attachment runs so deep, and is so hard to break, that a person cannot effectively make these changes even when they know they should.

This is where biology comes in. If it becomes possible to safely target the underlying neurochemistry that supports romantic attachment, using drugs or other brain-level technologies, then there is reason to think this could help some people who really need it. At the very least, we claim, there is reason to look into this possibility given the stakes involved. And if it does turn out to be possible, what then? We would need to determine the ethical limits for using such drugs and decide what legal and policy structures would have to be put in place to guide their responsible use.

These challenges come to the fore in later chapters. But at no point do we advocate the use of biotechnology as a quick fix for relationship troubles. Nor do we think such technology should be used illegally, coercively, or at home in isolation. Instead, we consider the voluntary use of biochemical agents in conjunction with professional psychotherapy, social support, and other established strategies as a way to help people achieve their relationship goals.

Brave new love?

Even with these caveats, some readers may be grinding their teeth at the very idea of combining drugs and relationships (or maybe not; the title of this book is not subtle). The last thing we need, they might think, is another excuse for pharmaceutical companies, or the medical-industrial complex more generally, to colonize another aspect of our lives. Or maybe they see parallels with Aldous Huxley's fictional drug "soma" in the dystopian *Brave New World* (the original working title for our book was *Brave New Love*). As an online commenter wrote in response to a piece about our work: "Shall we call it denial in a bottle? This is a horrible idea, you can't fix everything with a happy pill. I thought we had already established this."

The commenter is making an important point. There are many legitimate concerns that could be raised about the romantic version of a happy pill, and we will look at those concerns and take them seriously. But it won't be enough to just catalogue objections and call it a day. Whenever society is faced with some new biochemical or technological possibility, we have to think carefully, not just about the potential costs and dangers of making it a reality but also about the potential benefits it could provide if used responsibly and managed effectively. Throughout this book, we will highlight cases where we think love-affecting biotechnology could be genuinely helpful in promoting people's welfare or reducing their unbearable suffering. We'll also spell out a number of boundary conditions for when such drugs should *not* be used for various reasons.

These drugs are not science fiction. In the case of Sofia, there are existing medications she could use off-label—that is, for something other than their intended purpose—to dull her feelings of attachment today. We have also mentioned MDMA, which might at first seem like a pro-love drug, or something that inevitably draws people closer together. For an abusive relationship, that could spell disaster. But as we will see, MDMA is not quite as simple as that. It isn't just an emotional glue that cements people's hearts together no matter how incompatible they may be. Sometimes it can help a person realize that a relationship needs to end, and that the sheer fear of change has been holding them back. As we mentioned, MDMA is now being tested as a treatment for individual-level mental health issues. Are the scientists behind these trials even asking about the drug's effects on relationships? With the exception of the Mithoefers, for the most part, no. But they should be.

This lesson applies to drugs used for medicine in general. Overwhelmingly, such drugs are used to treat problems faced by an individual, not problems that exist within a romantic partnership. Seeing the limitations of this approach, some psychiatrists have started to get creative with diagnostic categories. One of them, whom we'll meet in a later chapter, has devised a way to prescribe medication for the treatment of *jealousy*. This is not something that is normally seen as a medical condition, although unfounded or excessive jealousy can of course do serious damage to an otherwise good relationship.

There is some evidence that drug-based treatments for obsessive-compulsive disorder may also help with destructive jealousy (a fact the psychiatrist took careful note of). But it's all a bit touch and go. This book argues that we should study the implications not just of currently illicit drugs like MDMA, but also of common, legal, prescription medications that are already affecting our relationships—only in ways we don't fully recognize. This is a blind spot in Western medicine: the tendency to ignore the interpersonal effects of drug-based interventions. Hormonal birth control is a key example in this area; antidepressants and antianxiety drugs are as well. It should be a scandal that we don't know more about the effects of these drugs (good or bad) on our romantic partnerships, due to an exclusive focus on individuals and their private symptoms in clinical studies.

How can scientific research norms be overhauled to take *relationships* more fully into account?

The time to think through such questions is now. Biochemical interventions into love and relationships are not some far-off speculation. Our most intimate connections are already being influenced by drugs we ingest for other purposes. Controlled studies are already underway to see whether artificial brain chemicals might enhance the positive effects of couples therapy. And as later chapters will explore, fundamentalist religious groups are already experimenting with certain medications to quash romantic desires—even the urge to masturbate—among children and vulnerable sexual minorities.

Simply put, the horse has bolted. Where it runs is up to us. Some may fear the advent of relationship-affecting neurotechnologies and do everything they can to prevent their further development. Others may think that in the future doctors will have a moral responsibility to chemically help patients with relationships and love. Others may fall somewhere in the middle.

Love and its effects touch all of us. So each of us must decide where we stand in this debate. The goal of this book is to arm you with the latest scientific knowledge and a set of ethical tools you can use to decide for yourself whether love drugs—or anti-love drugs—should be a part of our society. Or whether a chemical romance might be right for you.

CHAPTER 2

LOVE'S DIMENSIONS

AMONG THE YUSUFZAI PUKHTUN, a tribe in northern Pakistan, it is rumored that the most powerful love potion you can get is water that's been used to wash the body of a dead leatherworker. In Swedish folklore, to capture your crush's attention, you should carry an apple in your armpit for a day—and then present it, bathed in your own special scent, at an opportune romantic moment. Since Roman times, at least, a long list of weird tinctures and funny foodstuffs have been thought to stimulate lust, love, and good relationships. Sometimes love doesn't just happen. Sometimes you have to help it along.

Anti-love remedies have a fabled past as well. According to one ancient tradition, a little bit of bloodletting would cool your passions. So would drinking plenty of water, getting lots of exercise, and avoiding rich food and wine. The Roman poet Ovid, in his *Remedia amoris* (The cure for love), recommends taking on multiple lovers to redirect your erotic desires, distracting yourself with friends or fighting in a war (if you can find one), and avoiding sappy poetry. He also warned against turning to "harmful herbs" and the "magic arts." Such potions, he counseled, were a sham.

And they *were* a sham—or wishful thinking, to put it less harshly. But today, such fantastical visions are becoming technological reality.

Thanks to recent advances, not in witchcraft or wizardry but in neuroscience and related fields, the biological underpinnings of romantic love are being revealed. Some scientists think that the more we understand this biology, the more we will be able to influence or even manipulate those factors directly, through biochemical intervention. In other words, the prospect of real-life love drugs (and anti-love drugs) is now upon us.

♥ ♥ ♥

It is difficult to talk about love—romantic love, anyway, which is the kind we have in mind for this book—without immediately stumbling into clichés. Love is discussed so much that it can seem as though there is nothing new to say about it. Even when it *is* groan-inducing, being neck-deep in love can be one of the best experiences of life, and being bereft of love can be one of the worst. This is why love, the hope of love, or the loss of love tends to inspire dramatic movie plots, emotional songwriting, and poetry (even by those who would never be caught dead scribbling rhyming couplets otherwise).

Love also has its share of downsides. Friendships are often thrown under the bus in the race for romance. When we're in love, we want to stay in love, and this can sometimes feel like a high-stakes battle for survival. It can make us overly protective, jealous, and anxious. When we lose love, this tends to be among our most profound losses, and we may obsess about what we could have done differently. Unrequited love can be torture. And missed opportunities for romance that are only fully recognized later may fuel deep regret.

The two of us have thought a lot about love. We've each been in love (more than once; not with each other). We've experienced heartbreak. We've written our share of bad poetry. But as philosophers and bioethicists, we have also tried to get a grasp on different ways of understanding what love is—and on the practical and ethical significance of intervening, literally, in the "chemistry" between romantic partners.

Love and other drugs

Enter love drugs, two words that don't normally go together, and each one meaning something different to different people. Before diving into the science and ethics of love-enhancing biotechnology, then, we need to say what we mean by these terms.

We'll start with "drugs" because the concept is more straightforward. Drugs are just chemicals. That is essentially the entire definition. In practice drugs are usually thought of as specific chemicals that can be relatively easily taken into the body, whether by swallowing, snorting, inhaling, or absorbing them through the skin, which have some kind of physiological or psychological effect.

The *kind* of effect usually matters for how we regard these chemicals, though how this matters is not straightforward. When the intended effect is to cure or reduce the symptoms of a disease, drugs are usually called medicine. When the intended effect is recreation, spiritual development, or the exploration of alternative states of consciousness, drugs are usually not called medicine and just keep the generic label "drugs." In fact, one and the same chemical substance might be considered "medicine" in one context, if used in a particular way or toward a particular end, and just a regular old "drug" in another context, if used in a different way or toward a different end. Even *within* the category of medicine, a single drug might produce the intended effect or a side effect depending on why it was prescribed. A popular textbook on pharmacology makes the point like this:

> Amphetamine-like drugs produce alertness and insomnia, increased heart rate, and decreased appetite. Drugs in this class reduce the occurrence of spontaneous sleep episodes characteristic of the disorder called *narcolepsy*, but they produce anorexia (loss of appetite) as the primary side effect. In contrast, the same drug may be used as a prescription diet control in weight-reduction programs. In such cases, insomnia and hyperactivity are frequently disturbing side effects. Thus therapeutic and side effects can change, depending on the desired outcome.

We will say more about drugs later on, focusing on this issue of classification. As we will see, the decision to call a chemical substance a "drug" or "medicine" has important social and ethical implications. But before we get into that issue, we have a much more difficult task before us, which is to clarify what we mean by "love."

♥ ♥ ♥

In the previous chapter, we explained that we will often use the word "love" informally, especially when quoting people or referring to their experiences as they understand them. In those situations, what we mean by love is whatever makes sense in the context, and this will usually be pretty clear. In other situations, we will bring up core features of love that tend to appear in a range of formal definitions, usually to give an account of how love *in that sense* could be affected by a chemical agent. But we don't come down in favor of a single definition to use throughout the book. Mainly, this is because we don't want our analysis of particular cases to depend on which theory of love you happen to agree with.

That being said, we do think that any plausible theory of love would recognize that it has, at minimum, a dual nature: two fundamental aspects that go together to make love what it is, neither of which can be ignored. In this respect we agree with the philosopher Carrie Jenkins, who has recently defended a "dual nature" account in her excellent book *What Love Is*. As Jenkins argues, the concept we are after cannot simply pick out a biological phenomenon, as in theories that reduce love to some kind of animalistic drive; but nor can it simply refer to a social or psychological construct or something that exists in a disembodied soul. Although you may have heard that romantic love was invented in the West in the last few hundred years, it wasn't. It has been around (in one manifestation or another) since the dawn of our species, ingrained in our very nature. But the particular *forms* it has taken—as a result of the diverse ways people have understood it, reacted to it, molded it, and tried to control it or set it free—have indeed been different in different places and throughout different periods of history.

The idea is simple. Love would not exist as we know it if we did not have certain built-in biological drives related to attachment and mating. Moreover, the underlying function and makeup of our biology puts certain constraints on how we experience and even think about love at a higher level. At the same time, beliefs, norms, and expectations about love vary from culture to culture and may change over time; these higher-level factors can also affect our experiences and conceptions of love. They can even alter certain aspects of our neurochemistry. (If you don't see how something like beliefs could affect neurochemistry, just imagine you are Oedipus Rex and you've recently learned that your lover Jocasta is your mother. You can bet this belief will put a damper on your sex drive.)

In contemporary Western society, three main clusters of beliefs about love tend to show up on the psychosocial side. These are the concepts and representations of love that appear in art, literature, pop culture, and everyday discussions. In no special order, we have (1) the idea that lovers should be a "good match" or "made for each other" (as in the notion of soul mates); (2) the idea that lovers should value each other for who they are in particular, for what makes them distinctive and irreplaceable; and (3) the idea that lovers should have a steadfast commitment to each other—usually a sexually monogamous one. We will weave in and out of this psychosocial dimension as we go along. First, though, we'll give a quick overview of the biological dimension of romantic love.

In rough outline, the science goes like this. Underlying love is a set of overlapping but functionally distinct brain systems that evolved to suit the reproductive needs of our ancestors. These have been described differently by different theorists, but the most prominent account breaks things down into three systems: the *lust* system, the *attraction* system, and the *attachment* system. The role of lust or libido—as assigned by natural selection—is to inspire interest in a range of potential mating partners. The attraction system then narrows our focus down to a smaller number of partners, often one in particular. And the attachment system supports the formation of a long-term pair bond, which would have been important for successful childrearing in our ancestral environment.

Different brain chemicals, including testosterone, oxytocin, and dopamine, regulate these partially independent systems; the actions and reactions of these chemicals are largely responsible for our interpersonal drives and emotions. According to some theorists, these chemicals and the neural pathways along which they travel form the universal building blocks of romantic love. These building blocks are then reflected in, as well as shaped by, the sociocultural factors that bear on love across time and geography.

So, although there is wide variation in both subjective experiences and conceptions of love from person to person, between cultures, and over time, the thought is that—biologically speaking—the same basic "machinery" is under the hood. And by tinkering with this machinery through the application of certain biotechnologies, we are suggesting that it should be possible to influence those aspects of love that manifest "above the hood."

Car analogies should be used sparingly. But let's ride with this one a little longer. Obviously, the way a car runs, including how and where it moves through space, is not just a matter of internal mechanical aspects (corresponding to brains and biology in this analogy), like the rate at which pistons fire or how fuel is moved through the engine. It's also shaped by external factors, like the beliefs and decisions of the driver (individual psychology), as well as the presence or absence of pedestrians, the commands of traffic signals, and arbitrary, which-side-of-the-road conventions (sociocultural norms and physical environment). In the same way, the course and character of love is not just a matter of neurochemicals, genes, and so on. Instead, what love *is* in a given context is constrained and informed by a complex set of outside forces that derive from history and society and interact with individual minds and behavior.

These forces range from prevailing cultural norms and assumptions about love—the stuff of books, films, poems, pop songs, and TV shows—to the explicit categories and language people use to describe love, to how people make sense of their experiences of love in terms of those categories and norms. These outside forces, too, are subject to change.

Tinkering with biology, then, is not the only way to modify love: its psychosocial aspects can be tinkered with as well. At a societal level, people might try to challenge existing narratives about love, including dominant norms for how love should manifest in different relationships. Should love require sex and passion, for example, to count as truly romantic? Or is romantic love more about loyalty and working through difficult problems? Different societies, or the same society over time, might emphasize different factors.

As these norms and narratives change, so too will the psychosocial side of love, including what counts as love in a given social context. At a more personal level, too, it may be possible to change how one experiences and even conceives of love by adopting different attitudes, changing one's behavior or circumstances, or committing to an alternative lifestyle with its own set of values for conducting intimate relationships. (We will look at some examples of these kinds of changes in the pages to come.)

The important point for now is that social, psychological, and wider historical factors cannot be discounted. As the American psychologist and feminist Lisa Diamond argues: "Calling attention to the biological substrates of love and desire [does not] imply that biological factors are more important than cultural factors in shaping these experiences. On the contrary, research across many disciplines has shown that human experiences of sexual arousal and romantic love are always mediated by social, cultural, and interpersonal contexts, and ignoring these contexts produces a distorted account of human experience."

At the same time, she continues, ignoring the biological underpinnings of human psychology and romantic behavior produces an equally distorted account: "human sexual and affectional experiences are neither 'mainly cultural' nor 'mainly biological' but must always be understood as products of powerful interactions between biological and social factors."

In short, love has a dual nature. It is *both* biological and psychosocial, and it can be modified along either dimension.

From cars to art

To understand this dual nature, and to see how changes made along one dimension might manifest in the other, we'll switch from cars to art. Consider the *Mona Lisa*. This painting exists—*as* the *Mona Lisa*—by virtue of at least two dimensions: a physical, objective dimension that has to do with certain properties of the paint and canvas; and a more abstract, intangible dimension that has to do with historical background, subjective perception, and artistic interpretation.

Both dimensions matter. Yes, there is a sense in which a technically complete description of the *Mona Lisa* could be given in terms of splotches of paint occurring at various coordinates on a two-dimensional canvas. But most people would say that a really robust understanding of the *Mona Lisa* requires a range of more abstract levels of analysis: how the piece appears to the person looking at it, something about the emotion expressed by the subject's peculiar smile, what we know about the artist who made the painting, the cultural context of its production, and so on.

It is tempting to think that such nonphysical attributes are the ones that really count. In the analogy with love, you might think that what is happening in your brain is nothing compared to what it actually feels like to be in love, for example, or to the role love plays in your life. But the subjective and objective dimensions are not so easy to keep apart. Brain chemistry, like the splotches of paint that constitute the *Mona Lisa*, is fundamentally a part of what love is. Just imagine that someone got their hands on the *Mona Lisa* and proceeded to scrape off all the green splotches of paint and replace them with shades of hot pink. This physical change would radically alter the *Mona Lisa*, not just the two-dimensional canvas. You can fill in the love part of that analogy on your own.

Finally, consider the issue of artistic restoration. As masterpieces like the *Mona Lisa* fade with time, they can sometimes be touched up to maintain their essential character. Just as with the green-to-pink transformation, this sort of alteration would happen by making physical changes to those splotches of paint (although no hot pink should be used for restoration). Of course, different theories of art or different restoring artists might disagree about what is really essential to the painting, and therefore about which specific changes are needed to preserve its vital features. But there will be a limited range of plausible interpretations: the historical and subjectively perceivable aspects of the painting, plus the objective properties of the paint and canvas, will rule out certain approaches to restoration.

You can see how a similar analysis might apply to the biochemical "restoration" of romantic love. Different theories of love will emphasize different features as being essential, whether in general or as applied to a particular couple; but both psychosocial and biological factors, considered in a given context, will constrain the range of interventions that could plausibly count as love-restoring. In the next chapter, we will consider some objections to this idea—the idea that love itself could be changed or preserved through biochemical manipulation—but let's give ourselves a head start in fending off those objections by turning now to another analogy. This one comes from Carrie Jenkins, and it is designed to make love's dual nature even more intuitive.

The biological actor and the social script

Imagine you are watching *Star Trek* and you point to the screen and say, "There's William Shatner!" And then you point to the screen a second time and say, "There's Captain Kirk!" Now, you could certainly be accused of strange behavior, but you are not contradicting yourself. What's on the screen just *is* William Shatner, embodying the role of Captain Kirk; or, it is Captain Kirk, embodied by William Shatner (take your pick). Similarly, from a dual-nature perspective, love just *is* a cluster of biological factors experienced subjectively and shaped by various societal constraints; or, it is a cluster of socially mediated subjective experiences, shaped by certain biological constraints (again, take your pick). It's both, not one or the other.

Thinking about love this way raises fascinating questions. For example, which aspects of love are best described at the psychosocial level, and which at the level of biology? Let's go back to *Star Trek* to tease this out. As Jenkins notes, when we hear the line "Beam us up, Scotty!" we know this comes from a script that was written for a character (Captain Kirk); but the mouth on the screen pronouncing that line is the mouth of an actor (William Shatner). Similarly, when it comes to love, we can ask which parts come from the prevailing social script, guiding us toward certain thoughts and behaviors; which parts come from the underlying biological processes; and how the two interact with each other in practice.

Consider the example of a lesbian couple in late nineteenth-century England. Biologically, Jenkins tells us, they are in love: the neural and hormonal mechanisms responsible for desire, attraction, and romantic attachment are all firing away. The couple is experiencing the rush of adrenaline, the high of serotonin and dopamine, the pull of oxytocin, and all the other felt consequences of the brain chemicals that are coursing through their biological systems. Given the historical context, however, the available social scripts to act on those feelings are highly restricted. These two women cannot get married. They cannot publicly express their affection without putting themselves at risk. They cannot raise children together. In short, the biological dimension of their love is disabled from manifesting in the social sphere as it would for a heterosexual couple.

In fact, their feelings for and commitment to one another—as passionate and sincere and deeply rooted as they are—might not be recognized as a true form of love by members of the wider society. This lack of recognition, in turn, could shape how they conceive of their own relationship, interpret their own emotions, and behave even when they are alone, all of which might affect what is happening biochemically between them. Feeling shame, for instance, can interfere with sexual desire or inhibit its expression, which can decrease sex hormone levels and curb the release of oxytocin. This in turn can erode attachment, which might affect desires and behavior, and round it goes. Simply put, historically contingent norms and expectations drastically limit this couple's ability to engage in—and have others recognize—the social aspects of the biological bonding mechanisms normally associated with romantic love.

Changing love's two natures

Things have improved for lesbian couples in many countries since the nineteenth century. The social script for love has in several respects dramatically changed. Political debates, philosophical arguments, appeals to people's sense of common decency and shared humanity; plays, books, movies, and television shows presenting alternative views of gay and lesbian relationships—all of this has combined to make same-sex love a coherent concept with wide recognition. So the social side of love can evidently change in response to concerted interventions. Moreover, it can do so in a way that many people (including the two of us) see as morally good.

What about the biological side of love: can it be changed? Suddenly, things grind to a halt. There is an immediate sense of danger here, and rightly so. When the social script is held constant and the biological "actor" cannot seem to pull off the role they have been assigned, societies have often tried to modify the actor, saying in essence, "The show must go on." We are thinking here of sexual orientation conversion efforts in homophobic settings (many of these efforts have been biologically based). There are also chastity belts, genital mutilations intended to curb sexual desire, teaching children they will go to hell if they masturbate, punishing people for falling in love with the "wrong" person, and so on. History is full of examples of societies trying to retrain recalcitrant actors, often in vain, in an attempt to force them into ill-suited roles. There are many examples still today.

Perhaps we should conclude, then, that the biological actor is right, at least most of the time, and if there is a mismatch of some sort, it is the social script that should be expected to change. We have defended something like this view before, calling it the principle of Default Natural Ethics (DNE). All else being equal, we argued, society should adopt scripts (like institutions and norms) that are maximally consistent with people's evolved biological natures—for example, their sexual orientations. In this way, society can avoid the bad effects of misguided behavioral prohibitions and repression, whether enforced through law or custom, on individual and collective well-being. This is how the great German American philosopher and social critic Erich Fromm expressed the idea in his 1947 treatise *Man for Himself: An Inquiry into the Psychology of Ethics*:

> If man were infinitely malleable then . . . norms and institutions unfavorable to human welfare would have a chance to mold man forever into their patterns without the possibility that intrinsic forces in man's nature would be mobilized and tend to change these patterns. Man would be only the puppet of social arrangements and not—as he has proved to be in history—an agent whose intrinsic properties react strenuously against the powerful pressure of unfavorable social and cultural patterns. In fact if man were nothing but the reflex of culture patterns no social order could be criticized or judged from the standpoint of man's welfare since there would be no concept of "man."

If you have seen William Shatner play certain roles besides Captain Kirk, you get the idea. Shatner, the actor, has only so much range, and if you try to squeeze him into a role he is not cut out for, the result can be—to use Fromm's word—"unfavorable." We humans are like William Shatner. Our biological nature places certain limitations on the range of social scripts we can follow while still pulling off a decent performance (or experiencing a minimum of flourishing). If you force people to act in ways that are too much at odds with their evolved dispositions, things will get ugly. (We have nothing against William Shatner, by the way.)

Sexual repression is a key example. It has been argued that puritanical sexual attitudes in the United States coupled with abstinence-only sex education policies have backfired, as seen in higher teen pregnancy rates and increased incidence of sexually transmitted infections among American religious youths. Even the moral crisis of child sexual abuse in the Catholic Church has been (controversially) linked by some scholars to norms of celibacy in the priesthood. These kinds of examples should give us pause. At a minimum, they should make us wonder whether widespread (or heavily institutionalized) suppression of the human sex drive—given our nature as a sexually reproducing species—might ultimately do more harm than good.

According to the principle of DNE, it is only when biologically friendly norms about human drives and behavior lead to harm or violate clear moral standards, like justice, that laissez-faire policies should be reconsidered. Then the costs and benefits of different forms of social regulation can be tallied up. Consider the example of forced sex. This occurs throughout the animal kingdom and in humans, it is a form of rape (some forms of rape do not involve physical force, which is why we are drawing this distinction). Some researchers have described such behavior as being ultimately rooted in biology: it appears to be employed by our primate relatives and may have been a way for low-status ancestral human males to pass on their genes despite not being able to find a willing mate. According to this way of thinking, rape is in some sense a "natural" phenomenon.

Obviously, this perspective is controversial, not just morally or politically but also scientifically; the evidence raised in support of this theory has been sharply criticized. (Among other problems, it shifts attention away from sociocultural norms and structures that tend to be permissive of male violence in human societies, when these are central to understanding the more immediate causes of rape in the contemporary world.) The point we are making is more conceptual. It is that *even if* rape could be meaningfully explained in terms of biological factors (in addition to social, psychological, and cultural ones), this could never ethically justify rape. Rape is wrong not because it is natural or unnatural, biological or social, or motived by a desire for sex or power. It is wrong because it is a gross violation of another's sexual autonomy and bodily integrity and often causes grievous harm. It is on that basis that we should condemn sexual assault in all its forms, punish it to the fullest extent of the law, and try to prevent it from happening whenever we can.

In a similar vein, some scientists believe that pedophilia—that is, primary erotic interest in pubescent or prepubescent children—is at least partially explainable in terms of biological factors: stable, internal factors present at birth that a person cannot typically choose (not unlike a sexual orientation). So, for example, the finding that males with pedophilia are more likely than other males to be left-handed and have low IQs has led some researchers to speculate that pedophilia may be causally related to neurodevelopment. In other words, associations might exist between left-handedness, low IQ, and pedophilia because a shared biological process underlies all three.

This perspective, too, is controversial. But let us assume for the sake of argument that these sorts of findings turn out to be widely replicated. Let us even assume that pedophilia is one day recognized as a kind of sexual orientation, albeit one based on age rather than sex or gender, as some researchers have proposed. Would this suggest that we should rewrite the social script condemning sexual contact with children—that is, acting on a pedophilic orientation—so that those with a biological disposition toward pedophilia could more readily express that aspect of their nature?

Of course not. Although we should not necessarily stigmatize those who have, but do not act on, such misdirected sexual preferences (especially if they consciously reject the preferences and are seeking help to control their behavior), we should uphold the strongest possible norms against sexual abuse of children. Moreover, if a person with pedophilia volunteered to *change* their biology in an effort to snuff out their dangerous desires—through the ingestion of testosterone blockers, for instance—some would say they should be commended.

In defense of the unnatural

The "naturalistic fallacy" is one of the oldest and most famous mistakes in ethics. It draws conclusions about what should exist or be done directly from descriptive statements about how things are, such as what our natural or biological dispositions happen to be. But something's being natural or biological does not by itself, logically, morally, or in any other way, entail that it is good or desirable. Cancer is natural but bad. Eyeglasses are unnatural but good. Naturalness and moral status can come apart. So while our claim is that societies should broadly favor cultural and ethical norms that align with people's deepest biological traits and dispositions, we must also be capable of morally improving ourselves, individually and as a species.

How does this apply to love? As we have seen, our capacity for love has a strong biological component. It is also heavily shaped by social norms and other outside pressures from the surrounding culture. Both of these dimensions can combine to create conditions for love that are bad for us or that could be seriously improved (as we will illustrate throughout this book). So we might sometimes have a reason to *intervene* in one or both dimensions to help love reach its full potential.

Consider the parallel case of happiness. Happiness is, by most accounts, something that is intrinsically valuable—just as love is. Moreover, a number of environmental, social, psychological, and biological factors interact with each other in complex ways to influence whether, how, and to what extent we are actually happy. Finally, there is a low-level, common pathway for these various factors that can be analyzed in terms of events in the brain. This pathway can be modified directly, using certain drugs for instance, or indirectly, by changing one's thoughts, behavior, social context, physical environment, or any combination of the above. Depending on the specifics of a situation, it may be appropriate to make one or more such modifications, including the biological one (for example, with a drug like Prozac), in order to increase the chance of living a happy life.

The point holds generally. For any complex phenomenon that exists at the interface of human culture, psychology, and biology, we may sometimes have good reason to intervene—in one or more of those dimensions—if our aim is to promote well-being. But whether we should actually do this in any particular case is almost never a question of whether the phenomenon at hand is natural or unnatural. Instead, what typically matters is the *moral* status of the target of intervention (Is it intrinsically bad or likely to cause great harm? If not, does it really need to change?); its *malleability* (Is it relatively easy to change, if not along one dimension then along another?); the *means* available for changing it (Is it feasible to intervene—biologically, psychosocially, or otherwise? How reliable and effective are the different options?); and the likely *ramifications* of intervening (Is it possible to predict with any confidence that intervening will make things better? If so, by whose standards? Who will benefit and who will be harmed? What are the likely trade-offs implied by each of the alternatives?).

In any case, caution will be required. It is true that nature has great beauty, that it is stunningly intricate, and that when we try to tamper with it, we often make a mess of things. We often lack sufficient knowledge and power to make actual improvements to such complex systems. But nature also allows for great ugliness and suffering, much of which is preventable. Take the case of us humans. The basic blueprint for our bodies, including our brains, was sketched out by natural selection—not to promote our flourishing in the modern world, but rather to ensure the survival and reproduction of our distant ancestors. And yet, our contemporary needs may be very different from those of our ancestors, and we may rightly value all sorts of things apart from passing on our genes.

We have to lose the idea that natural = good, and unnatural = bad. Some people have objected to medical treatment for deadly diseases—and even basic pain relief—as these are not found in nature (and thus presumed to be in some way bad). But people today enjoy longer lives, and in many respects far better ones, than ever before. Much of this improvement is due to genuinely beneficial modifications of natural phenomena through advances in science, medicine, and technology. (Much of it is also due to advances in ethics.)

In short, without knowing the specifics of a situation, it would be irresponsible to simply rule out potential changes to biology as automatically unjustified or playing God. Nuanced ethical judgments have to be made, taking everything into account. Sometimes, failing to intervene biologically will be morally wrong (just think about leaving a patient to die when you could have provided a medication to save their life). In other cases, there may be an "unfavorable" tension, in Fromm's sense, between the biological and the psychosocial, such that making changes along either or both dimensions could be appropriate or justified.

Resolving tensions

What kind of unfavorable tension are we talking about? Take an example that has nothing, at least directly, to do with love. Consider the situation of a person whose stable and deeply rooted gender identity (such as man, woman, or genderqueer) does not align with the gender role that was assigned to them at birth (typically based on their external reproductive characteristics, like whether they have a penis or a vulva). This experience of misalignment, which often has both biological and social components, can be profoundly distressing, to the point where something has to give.

A big part of the problem is that the social norms surrounding gender in Western society are so binary and restrictive. According to these norms, you are either a man (who must look, dress, and behave in particular ways) or a woman (who must look, dress, and behave in other ways), or you face the consequences. In our view, these norms should be changed and expanded and in general made a lot more fluid. In other words, a change to the social side is not only justified in this case, but sorely needed.

In the meantime, however, many people, including some who identify as transgender, feel that they must conform to one binary role or another, just to get by. (They have often been subjected to violence for failing to do so.) Others are more openly transgressive about gender norms and happy with that choice; still others feel that dominant gender norms and expectations suit them perfectly fine.

But let's consider a case where there is a clear tension between a person's biology and the relevant social norms. This will serve as a rough model for similar tensions when we come back to the case of love in the next chapter.

Imagine that you are a woman, but based on your outward appearance and the assumptions people make about you, you are regularly treated like a man. The philosopher Lori Watson has written eloquently about this exact experience from her own perspective, so we'll share some of her story to illustrate. As it happens, Watson does not identify as transgender. She was born with female reproductive organs, both internal and external, was raised as a girl, and has always identified with the female gender role assigned to her at birth. She is also nearly six feet tall, has broad shoulders, keeps her hair short, and happens to feel most comfortable wearing clothes that are typically associated with the male gender role in Western society. About 90 percent of the time, she says, she is mistaken for a man by strangers. This constant misgendering makes her daily life challenging in a way that many transgender people will instantly recognize. She writes:

> Living in the world as it is, I have to fight for recognition as a woman on a daily basis. My choices are to correct people when they call me "sir" or assume I am a man, or let it go, which often means functioning socially as a man. . . . Doing something as basic as going to the bathroom, anywhere that is public, is a nightmare. Few places have "unisex" bathrooms. So here are my choices: go into the women's bathroom and face public shouting, alarm, ridicule, and confrontation. Or go into the men's bathroom, look down at the floor, walk quickly into a stall, and hope no one pays any attention to me but face the serious fear that they might.

Now imagine confronting this sort of thing every single day, in countless ways and interactions. Imagine standing over the coffin of your grandfather, she says, and hearing the funeral director say to your father, "This must be your son I have heard so much about." Imagine your father replying, "No, this is my daughter." Imagine going to the emergency room for "what you believe to be an ovarian cyst, and the physicians and nurses are so baffled they ask you your name and repeatedly check your chart for your sex identification and then ask to see your driver's license, something you've already given at check-in, to triple confirm, one presumes, that you aren't delusional."

"I could go on," Watson writes. "I have hundreds of such stories. All this because your body is socially interpreted as masculine, yet you identify as a woman. These are my stories, and I was born female and assigned to the female sex, and I identify as a woman."

Given these kinds of experiences, it should be little wonder that some transgender people—who do *not* identify with the gender that typically corresponds to their sex assignment—feel that changing social norms is not enough. In order to simply survive in the world, or in some cases to manifest their gender identity in a way that is most authentic for them, changing biological features through hormone treatments or surgeries is the best option for some transgender people. And they may legitimately pursue these means even as they work with others to rewrite the constricting social scripts that are currently available for enacting gender.

What this example shows is that even if you think the social dimension of a phenomenon has serious problems—in this case, the problem of overly narrow expectations surrounding sex and gender, and the way these expectations are enforced in subtle and not-so-subtle ways—it does not automatically follow that the biological dimension of the phenomenon must at all costs remain undisturbed. Certainly, no one should be forced to undergo a biological modification to resolve such tensions, but some people might reasonably decide to modify aspects of their own biology, taking everything into consideration, *even if* they think the social script has certain flaws.

To summarize, how well our lives go is largely determined by a set of partially dissociable factors: the natural or social environment, cultural scripts or norms, psychology, and biology. Each of these can typically be modified in a number of ways. Absent a clear overriding reason not to, we should at least consider the full range of potential modifications and then choose the modification, or combination, that is most effective at improving our lives.

The same applies to human relationships. There may be biological obstacles to lasting love for many couples, especially in our modern societies. Intervening in this biology will sometimes be the best way to help a couple have a good relationship. In other situations, there will be strong reasons to prefer purely psychosocial interventions: they might be safer, more likely to be successful, or more fair or just (on the basis of limited resources, for example). But we shouldn't dismiss biological interventions from the start. Indeed, in some cases it may be the biological interventions that are safer, more likely to be successful, and demanded by justice. Or it may be some combination of both.

Absolutely none of this can be decided in the abstract, however. We need to have a concrete sense of the "unfavorable" tensions that actually arise in romantic relationships in contemporary societies, and how these tensions might relate to our biology.

CHAPTER 3

HUMAN NATURES

HOW MIGHT BIOLOGY AND SOCIAL FACTORS conflict in modern relationships? Well, it depends on the relationship. Is it monogamous? Polygamous? Polyamorous? Short term? Long term? What are the surrounding cultural expectations? What are the values of the partners? In this chapter, we will walk through some scenarios to show how tensions can arise between the biological actor and the psychosocial script in the sense we have discussed. We will start with the most common relationship model for long-term romantic partnerships in Western societies—namely, monogamy—and go from there to less common types.

Consider a married couple that has decided to be monogamous. As part of their wedding vows, they agree to be sexually exclusive with each other, so long as they both shall live. Most couples who get married make this promise. In fact, it's taken for granted in the prevailing social script for long-term relationships in Western (and many other) societies. But is this a good script? A plausible answer is, it depends—on the community, the couple, their beliefs and values, the wider context, and many other factors. And at least in part, it depends on the motivations behind the decision to be monogamous.

Some philosophers argue that monogamy can be morally good and valuable for a couple when it is motivated by a reasonable belief that it will fulfill the emotional and sexual needs of both partners. In other words, when sacrificing other sexual opportunities—for the sake of the relationship—is likely to promote the deep flourishing of both. On the other hand, monogamy that is motivated by jealousy, insecurity, or a desire for control will tend to have negative value.

Similarly, rejection of monogamy can be good or valuable when it demonstrates maturity and a well-grounded belief that the partnership is strong enough to thrive even as the partners engage sexually with other people. And rejection of monogamy can be negative when the motive is selfish or based in an attempt to avoid working together to address whatever sexual or other needs may be going unfulfilled.

It isn't just philosophers who make this argument. Dan Savage, one of America's leading sex-advice columnists, advances a similar view in his popular books and articles. The value of monogamy, he thinks, is not straightforward, nor is it necessarily universal. In other words, it may have different value for different people.

Savage is known for coining the term "monogamish" to refer to long-term, committed partnerships in which sex is permitted outside the relationship with certain conditions. This is a kind of compromise between monogamy and nonmonogamy that acknowledges the benefits and drawbacks of both. Some couples may find that such an arrangement works best for them. But even Savage does not reject full-fledged monogamy if it suits the needs of a given couple and they agree to it consciously and in good faith. It's just that, in his view, these conditions are rarely met. Instead, for most couples, and for society at large, the discourse surrounding monogamy and human sexuality is often deeply dishonest.

Some people "need more than one partner," Savage writes, "just as some people need flirting, others need to be whipped, others need lovers of both sexes." He argues that we cannot normally help our innermost desires, nor should we lie to our partners about them.

We need to face up to reality. "In some marriages, talking honestly about our needs will forestall or obviate affairs; in other marriages, the conversation may lead to an affair, but with permission. In both cases, honesty is the best policy."

Honesty allows people to build their relationships on an authentic foundation: on whatever is real for each individual, and real for both of them together. Not on sugar-coated platitudes. "I acknowledge the advantages of monogamy," Savage writes, "when it comes to sexual safety, infections, emotional safety, paternity assurances. But people in monogamous relationships have to be willing to meet me a quarter of the way and acknowledge the drawbacks of monogamy around boredom, despair, lack of variety, sexual death and being taken for granted."

Is monogamy really all that bad? Some people argue that monogamy is actually natural for humans; if it seems to have downsides, that is a weakness in the person or couple, not monogamy itself. Or in a more structural version of this perspective, it's the fault of society, post–sexual revolution—a society whose moral laxity around sex and relationships we should regret and try to resist. As one religious scholar has put it, we should "develop laws, policies, and curricula to teach the basics about the [monogamous] nature of human sex and marriage, and to encourage and facilitate citizens to live their sexual lives in accordance with the natural norms and limits that govern us all."

There are two main problems with this argument. First, it isn't clear that monogamy *is* natural for humans. And second, even if it were natural—in some sense—this wouldn't automatically entail that a monogamous relationship norm is something we should all adopt, much less enforce across the board. As we learned earlier, what is "natural" and what is good can sometimes part ways.

Let's take the first issue first. Is monogamy natural for our species? As we proposed with our DNE principle, the default move, which can be overridden as necessary, should be to adopt and promote social scripts that are at least compatible with those aspects of human nature that do not respond very well to cultural suppression (in the sense spelled out by Erich Fromm). If that much is right, then the question of whether we should change the current social script for love—the one that frames monogamy as the relationship ideal to the exclusion or denigration of other contenders—depends in part on facts about human nature as they bear on sexual exclusivity.

Here is what we know. Humans, as a species, are generally disposed to form at least one long-term pair bond with another person, with whom they will often raise children. This disposition to develop a close emotional attachment, as a precursor to parenthood, is part of our biological makeup (though it does not apply to every individual). Yet very few nonhuman animals are sexually exclusive in mating relationships, even in the context of pair-bonded attachments. From an evolutionary perspective, such stringent fidelity tends not to maximize reproductive success, which means it's unlikely to survive as a dominant mating strategy.

Humans are no exception. Quite a lot of sex occurs in our species outside of primary, committed relationships—as everyone knows—and this is true even in societies where such behavior is strongly condemned and heavily policed. As a general rule, if some behavioral tendency is pursued at great risk or comes with a very high social cost and yet remains widespread across cultural variation, its roots are likely grounded in biology.

Okay, then, monogamy is *not* natural for our species. Is that right? In fact, that is not the full story either. There are two senses of "monogamy" that tend to get conflated in discussions like this, and teasing them apart will help to resolve the puzzle. According to the prominent evolutionary theorists David Barash and Judith Eve Lipton, a crucial distinction needs to be drawn between *social* monogamy and *sexual* monogamy. Two animals (of any species) are socially monogamous if they live, nest, forage, and copulate together over an extended period of time, often working together to rear offspring—like most humans. But social monogamy does not always entail sexual monogamy, even "out there" in nature as biologists had long assumed. Thanks to

DNA fingerprinting and other modern techniques, we now know that many nonhuman animals sleep around (as we might say) much more often than was previously believed. Even birds, famous for their social monogamy, have been unmasked as sexually adventurous by studies looking at the genetics of their offspring.

So much for sexual monogamy in most of the animal kingdom. Yet even social monogamy turns out to be relatively rare among nonhuman animals, including our closest primate relatives. Where does this leave *Homo sapiens*? Scientists are not actually sure about just how wide the gap is between (social) appearance and (sexual) reality when it comes to monogamy in our species. There is disagreement about whether we humans are (1) mostly monogamous, both socially and sexually, with a little bit of something extra on the side (closer to the bird end of the spectrum), or (2) mostly socially monogamous but sexually nonmonogamous, so that claims of "natural" sexual monogamy are little more than agitprop for outdated and oppressive cultural norms.

Something like the second view has been defended by sex researchers Christopher Ryan and Cacilda Jethá in their popular book *Sex at Dawn*. "Like bonobos and chimps," they write, "we are the randy descendants of hypersexual ancestors." Accordingly, "Conventional notions of monogamous, till-death-do-us-part marriage strain under the dead weight of a false narrative that insists we're something else." The campaign to obscure the "true nature of our species' sexuality leaves half our marriages collapsing under an unstoppable tide of swirling sexual frustration, libido-killing boredom, impulsive betrayal, dysfunction, confusion, and shame."

If Ryan and Jethá are right, the prevailing social script that valorizes sexual monogamy probably should be revised. But not everyone is on board with their broad-brushed claims, including other leading sex researchers. For a different perspective, we reached out to the evolutionary psychologist Geoffrey Miller, who specializes in human mating.

"Everybody who studies this topic," he told us, "knows that we tend to engage in lots of nonmonogamous behavior, including adolescent sexual experimentation, short-term casual mating during young adulthood, occasional affairs even given pair bonds, mate-switching to new pair bonds (serial monogamy), and polygyny among high-status males."

But this doesn't mean we are just suit-sporting bonobos. According to Miller, the real disagreement among sex researchers is whether most people most of the time are disposed to be in long-term pair bonds that are highly sexually exclusive, or whether most people most of the time are disposed to be in (one or more) long-term pair bonds that are relatively open to other relationships. As he sees it, the "scientifically serious" spectrum of opinion ranges from "whether we're about 80% monogamish," in Savage's sense, "to whether we're only about 50% monogamish."

Carrie Jenkins, the philosopher we met in the last chapter, takes a similar view. Human nature is not homogeneous, she reminds us, nor are human mating strategies all of a piece. Indeed, Jenkins herself is drawn to polyamory: she has both a husband and a boyfriend, as she discusses in the prologue to her book. But even she thinks that Ryan and Jethá go too far. In her view, it is almost as if the traditional moralists—the ones who claim that human beings are monogamous (in both sense) by nature, and should act accordingly—have asserted that "blue is the natural eye color for humans, and then Ryan and Jethá, noting that a lot of humans have green eyes, [counter] that green is the natural eye color for humans." But why should it be one or the other? For some people, Jenkins points out, "monogamy really doesn't seem to be a terrible struggle. They say it feels perfectly 'natural' and delightful and right."

And for them it very well might be. But we need to be careful. What is natural for our species can be maddeningly hard to disentangle from deep-seated cultural expectations and psychological training. It is quite possible to feel that something is "natural" when really it's been drilled into our heads through oppressive socialization from when we were young.

According to Geoffrey Miller, even if we could magically filter out all heavy-handed social pressures, we would still likely find a lot of individual variation down at the level of "pure" biology. "Many people really are naturally very monogamous after their mid-20s," he told us, "forming happy long-term pair bonds with low probability of cheating or divorce." Other people, especially those who score high on a personality trait called sociosexuality, are "more naturally sex-positive, promiscuous, open to casual sex, interested in polyamory, and so on. Neither kind of person understands the other very well."

You can look at it this way. Just as some people may be "wired up" to be attracted to same-sex partners (we realize this is a gross oversimplification, and we hope you can forgive the electronics metaphor), other people may be wired up to feel attracted to, and desire a physical and emotional relationship with, more than one person at a time. The exact percentage of humans this applies to is not clear, and we are not really sure it matters. If homosexuality is natural for some people—that is, most consistent with their unchosen, innermost, most stable, hard-to-ignore preferences and desires—then polyamory is probably natural for some people, too, just as heterosexuality or monogamy may be for others.

Individual differences matter. For some people, the desire to have one main partner and to be sexually exclusive tends to be strongest and most stable, winning out over competing desires for multiple partners or sex with more than one person. People with this disposition most likely do well in societies with monogamy-reinforcing norms. For other people, the desire to have more than one partner, or sex with more than one person, is strongest and most stable, leading either to cheating (in the case of long-term socially monogamous arrangements), short-term serial partnerships (with or without a promise of fidelity), or, increasingly, attempts to enter into consensually nonmonogamous relationships, even if that conflicts with the dominant social standard. Of course, these wrestling impulses crosscut the dimension(s) of sexual orientation, affecting gay, straight, and bisexual people, as well as people who reject such simplistic labels altogether.

What all of this amounts to is that if we want a society where everyone, or even just most people, can really flourish in their romantic lives, we should push for a dominant social script that recognizes and allows for a range of relationship norms, so long as these are based on mutual consent and respect for others. That way, people can figure out what works for them, and be socially supported in their decisions.

"Every mating system requires years of practice to do well and ethically," Miller told us. That means years of social instruction and, where appropriate, sanctions to reinforce good behavior within the logic of each kind of system. In other words, while monogamy may be natural for some people, "it still takes a decade of relationship experience after puberty to learn how to have a happy pair bond." And while polyamory may be natural for others, it too "requires years of practice. None of these mating instincts arrive fully formed at birth, any more than our language instinct does." Yet for the right kind of practice to be possible for more than one relationship model, there needs to be greater awareness of ethical alternatives to social and sexual monogamy. As empirical studies are beginning to suggest, consensually nonmonogamous relationships can be just as functional and conducive to happiness as monogamous relationships, if properly supported. Such relationships should plausibly have a more prominent position in the social script for modern love.

Returning tensions

Now we come to the point of this chapter. Depending on your nature (in the sense we sketched above), your physical and social environment, and your core beliefs and values, various sources of tension can arise between these and related factors. This is what we were trying to illustrate with the example of gender-nonconforming people in the last chapter.

To see how this might play out in the context of love, suppose you are in a polyamorous relationship, one you have entered into thoughtfully and after much consideration. Your beliefs and values tell you this is the right sort of relationship, at least for you and your partners, reflecting your shared commitments to nonpossessiveness, sexual freedom, and openness and generosity about love. Chances are—in this day and age—that at least one source of tension for you is the existence of negative judgments from (some) members of the surrounding culture who see polyamory as selfish, sinful, or unsustainable. One way to address this tension would be to contest those negative judgments; difficult to do on your own, but if enough polyamorists came out of the closet and took a stand for their relationship values, they might make good progress on rewriting the social script. The gay rights movement once again provides a model.

Other sources of tension could be internal. Maybe you struggle to control your jealousy, despite believing that this emotion is not a reliable guide to moral judgments or good behavior. Jealousy has deep evolutionary roots. Most scientists think that it was an adaptive solution to the problem of paternity uncertainty in ancestral males and resource insecurity in ancestral females (the details are interesting, but we won't go into them here). In other words, the capacity to feel intensely jealous, especially when faced with a perceived sexual rival, is an ingrained feature of most people's relationship psychology.

Could a biological jealousy inhibitor help you meet your relationship goals? Something to supplement the more conventional methods of stamping out unproductive jealousy you might already be using (like therapy)? Consider the following real-life case study. It's about a monogamous couple, rather than a polyamorous one, but the principle remains the same:

> A 54-year-old male manager of an antique shop presented himself [to a psychiatrist] describing a depressed mood, intermittent sleep disturbance, as well as distinct appetite and weight loss. He volunteered two causes, a council election with one party's policies being a threat to his business, and recent marital dysfunction.

The "marital dysfunction" part caught our attention:

> Marital difficulties followed his stated wish to make contact with a former girlfriend, with whom there had been no contact for 30 years. His wife agreed pleasantly, however, a week later, asked if she could make contact with her old boyfriend. The patient progressively developed a set of anxiety symptoms including panic attacks, and depersonalization experiences. He began to harass his wife, initially about her motivation for re-establishing contact, and later pursuing all details of her original relationship with the man, questioning and haranguing her for hours. He accepted referral when his wife stated that his behavior threatened their 20-year-long happy marriage and that, if it persisted, she would leave him.

The man's psychiatrist dug deeper:

> Personality review suggested a man with a distinct obsessional personality: always on time for appointments, preoccupied with order and efficiency, extremely well organized, a meticulous checker of minor and major issues, perfectionistic, productivity-focused, scrupulous about morality and ethics, and distinctly conscientious. Symptom review indicated long-standing mild anxiety. A diagnosis of anxiety state with secondary depression was made initially, and he was prescribed alprazolam [Xanax] for a brief interval to reduce his anxiety. At the third consultation, the possibility of OCD [obsessive-compulsive disorder] was considered.

If you think about it, intense jealousy *is* a lot like OCD: fixation on tiny details, compulsive thinking, intrusive mental images, persistence with repetitive behavior despite clear signs it isn't helping (and may very well be making things worse). Perhaps the two phenomena have more in common than we realize. Researcher Paul E. Mullen has pointed out that "checking," an obsessive behavior associated with OCD, also occurs in people who experience jealousy: "an almost universal behavior among the jealous [is] checking that the lover is where they say they are and with whom they say, cross checking, re-checking (with frequent) inquisitorial cross-questioning of the lover."

Following this line of thinking, the man's psychiatrist changed the course of medication to clomipramine, sold under the brand name Anafranil, a common treatment for OCD. After six days of taking the drug, the antique-shop manager "stated that he felt less anxious . . . and that his jealousy (subjective feeling and associated behavioral actions) had eased considerably." A specialized program of exposure therapy and cognitive retraining was then started. Four weeks later, the man reported that the "original concerns were no longer relevant." His marriage was saved.

Part of what is interesting about this case is that jealousy is not usually seen as being some kind of mental disorder that's amenable to being treated with medicine. In fact, this man wasn't diagnosed with jealousy as such but was treated, in the words of his psychiatrist, "as if he had an obsessive-compulsive disorder." Emphasis on *as if.* What this goes to show is that it may not be necessary to pathologize—that is, characterize as a disease state—ordinary human experiences in order to recognize that our romantic biology can sometimes pull on our conscious thinking and behavior in unproductive ways, posing a serious threat to our well-being and that of our partners. Moreover, as we have just seen, sometimes a biological intervention, especially when combined with appropriate psychosocial or therapeutic approaches, can help eliminate that threat, whether or not we want to call the intervention "medicine."

Difficult choices

The last example was brought up in a discussion of polyamory, where jealousy of any kind is often seen as conflicting with the partners' highest values. But now suppose you are in a monogamous relationship. You and your partner genuinely believe that monogamy is the right way to go, all things considered: it is the closest fit between your natures, your social and cultural environment, and your mutual goals and values. In this case, instead of trying to tamp down jealousy about your partner's other partner(s), perhaps it is your own wandering eye that is standing between you and the happiest version of your committed relationship. Could science one day help you better stick to the social script you endorse and have agreed to follow?

Imagine you are in such a relationship. You and your partner are deeply in love. You are truly committed to each other, and you make each other happy in just about every way. But there is one thing gnawing at you, one persistent problem that you cannot seem to shake no matter what you do: your sex drive is much stronger than that of your partner. This asymmetry causes friction, despite your best attempts to dance around it. Your partner starts to feel guilty, even though you have promised it's no one's fault. And you start to feel resentful and unfulfilled, despite your best attempts to push your feelings aside.

Nothing is wrong with either of you. You just have different sexual needs—something that applies to many long-term relationships. Maybe you even started out with similar levels of desire and things shifted for one or both of you over time. There is nothing necessarily pathological about that either.

Now you have a life together. A house, kids, mutual friends. Neither of you wants to give this up. Your partner is just about perfect for you in every other way. You know sex isn't everything. You know that good relationships require sacrifice, and you are willing to do whatever it takes. But still you find yourself fantasizing about sex with others, and you fear that you will act against your values at some point if something doesn't change.

Now let's imagine that you and your partner have great communication. You have been honest with each other. You each know how the other feels. You've gone to a relationship counselor together and tried various exercises to sync up your sexual drives. No dice. Your own libido still rages, while your partner's stays calm and serene. (We'll talk about the prospect of boosting a low libido, like that of your partner, in a later chapter. For the purposes of this thought experiment, assume those methods haven't worked.)

In fact, you've even talked about opening up your relationship, trading a promise of monogamy for something more "monogamish." But only you would want to have sex outside of the relationship; your partner's needs are fulfilled within it. Uncomfortable double standards might develop. And anyway, it would be hard to explain to the kids. So you both conclude, after much discussion, that going this route would be more trouble than it's worth.

You seem to be facing a horrible choice. Either end a relationship that means the world to you, ignore or suppress your intense sexual desires, or end up cheating on your partner in a moment of weakness. What do you do?

Let's add one more detail to this story. Suppose you are taking some medication, prescribed by your doctor, for anxiety and mild depression—Bupropion, let's say (commonly sold as Wellbutrin). Unlike many antidepressant medications, Bupropion is not a selective serotonin reuptake inhibitor (SSRI), but rather a norepinephrine-dopamine reuptake inhibitor (NDRI), and it has fewer sexual side effects like loss of libido. According to one study, people treated with SSRI-based medications, compared to Bupropion, experienced "significantly decreased libido, arousal, duration of orgasm, and intensity of orgasm" relative to levels experienced before treatment.

The next time you see your doctor, you decide to ask about switching from an NDRI to an SSRI, knowing full well that this might lower your libido. Assuming that the latter is an equally valid treatment for your anxiety and depression, suppose your doctor agrees. You switch medications and you find that your desire for sex becomes more subdued. Your fantasies about affairs with others become less frequent. The friction in your relationship begins to subside. Once again, we have a case where a biochemical intervention could help resolve a persistent tension.

Taking stock

We designed that case to be intuitively sensible. You are supposed to think it would be justifiable, perhaps even praiseworthy, to switch your medication in order to save your relationship, keep your family intact, and adhere to your highest values. If you do think that, then we have cleared a low bar: it can *sometimes* be appropriate to intervene in love's biological side to help a couple achieve their relationship goals.

You might not agree with that perspective, however. You might think you shouldn't turn to medication, no matter what the cost to you or your family, to resolve tensions in your romantic relationship. Let us remind you that in this scenario you were already taking medication, and various strategies had been tried with no success. Even so, you might say, that is something you will just have to deal with; you should only change your medication if it's to address a medical problem, and desiring more sex than your partner is not a medical problem.

True enough, it isn't. There is nothing medically wrong with you if you happen to desire a lot of sex (just as there is nothing medically wrong with you if you happen to feel jealous about your wife's ex-boyfriend—within reason). In many cases, trying to find a partner with a similar sex drive, or one who is open to nonmonogamy, will be the best way for you to secure a happy relationship. Maybe even breaking up with your current partner—getting a divorce, splitting up the kids, moving to another part of town and starting over—is the price to pay for being true to who you are.

Or maybe you should stay in the relationship and try to convince your partner to reconsider your mutual promise of monogamy. Or else bury that part of who you are underground.

There is no one right way to proceed. The most justifiable course of action will depend on the details, and different couples will find that different solutions are (all things considered) best for them. It would be strange indeed if everyone had the same needs and values. But the best decision will be one that takes seriously the available options and weighs them against each other without prejudging the conclusion.

The balance of factors could work out different ways. At the very least, you should examine those factors rather than ignore them or take them for granted. If you are afraid to leave your relationship because you don't want to face the reality of your own sexual desires, for example, you should ask why that is. Are you ashamed? Have you internalized the idea that wanting sex is bad? What is the source of that idea, and is it reliable? Is the idea consistent with your other values? One thing is certain: staying in a relationship out of fear—fear of self-knowledge, fear of change, fear of disappointing your partner, fear of disapproval from society—is rarely a good long-term strategy.

At the same time, no one is reducible to their default sexual desires, however strong or mild those may be, and no one's happiness is reducible to the reliable fulfillment of those desires (though they usually are related). So it is not beyond the realm of possibility that you could have an all-things-considered good reason to stay with your current partner, even if this meant that some of your desires would go unfulfilled.

But suppose you've gone over all this a million times. You've considered every scenario, and you still aren't sure whether to make that big change, whether it's taking a relationship-enhancing drug (assuming that becomes legal) or ending your long-term relationship. If that describes your situation, we have some advice. Go for it. Make the change. You will probably be happier for it in the long run.

Evidence for this prediction comes from a recent study by Steven Levitt, the economist and coauthor of *Freakonomics*, who ran an experiment on a very large group of self-selected fence-sitters who were considering—between them—more than twenty thousand decisions. Participants went to a website and registered their dilemma. One choice was assigned heads, the other tails. And then the website would flip a virtual coin. If it came up heads, the site instructed the participant to go ahead and make the change. If it came up tails, it instructed them to preserve the status quo.

Participants could do whatever they wanted. Nobody was forcing them to obey a random website. But they did know what they were signing up for, and many of them took the plunge. Those who got heads were 26 percent more likely to make the change they'd registered compared to those who got tails, assuming honest reporting.

Up to six months later, the changers were on average much happier than the preservers, on the basis of surveys of both participants and third-party observers. Crucially, this effect was largest for life decisions that really mattered: decisions like whether to quit one's job or to break up with one's romantic partner.

In other words, the sheer fact that you are seriously considering a major change, up to and including altering your own biology with a drug or medication, might be evidence that you should make it. After all, if you were content with your situation, you wouldn't be awake all night contemplating the alternatives. But whether you make a choice to stay in a relationship or to leave it—with or without the help of a drug—you set in motion a new reality. Choosing *not* to make a change is still a choice, so choose with care. The status quo cannot not relieve you of this burden. Once we have the power to alter a situation, we are morally responsible for the decisions we make—including the decision to leave things to chance, or to keep things as they are.

At the end of the day there is no one-size-fits-all resolution to the various tensions that arise in romantic partnerships. We certainly haven't proposed one here. Instead, we've drawn on the dual-nature understanding of love, according to which love is fundamentally both biological and psychosocial. And we have suggested there may be room for worthwhile interventions along *both* dimensions if we want love and happiness to coincide in our relationships.

On the biological side, we looked at two possibilities—the use of common medicines to "cure" jealousy or to intentionally weaken a strong libido—but that is only the tip of the iceberg. Next we take a deeper dive into the currently available options.

CHAPTER 4

LITTLE HEART-SHAPED PILLS

IF YOU'VE BEEN ALIVE SOMETIME in the last thousand years, you will be familiar with the idea of a love potion—a magical liquid that when quaffed or applied makes a person fall passionately in love. In Shakespeare's *A Midsummer Night's Dream*, a love potion is introduced as a weapon in the marital war between the fairy king and queen, Oberon and Titania. To humiliate Titania, Oberon drizzles juice distilled from a "love-in-idleness" flower on to her eyelids as she sleeps, causing her to "madly dote" upon the first creature she sees—the donkey-headed Bottom. The flower's magical juice also causes a tumult for the play's mortal lovers; administered by the mischievous Puck, it makes Lysander and Demetrius both fall in love with the formerly spurned Helena, much to the dismay of Hermia, who was in the process of eloping with Lysander.

After many chases through the enchanted wood, all is set to rights, as this is after all a comedy. In Wagner's opera, *Tristan und Isolde*, a love potion is substituted for a death potion to avoid the demise of the romantic leads—but not until Act 3, as this is Wagner.

In both stories love potions force characters to assume emotions that are not their own. In many popular depictions, the effects of a love potion are often undesirable, even if the person deliberately seeks it out.

Consider the 1960s chart-topper "Love Potion No. 9" by the legendary songwriting duo Jerry Leiber and Mike Stoller. In this song, the narrator, a self-described "flop with chicks," visits Madame Ruth, who doses him with a black, turpentine-smelling potion brewed up in her kitchen sink. He's suddenly disoriented and starts impulsively smooching everything and everyone he sees. This includes, as you might remember, a police officer who subsequently smashes his precious "little bottle." Presumably, none of this improved the narrator's luck with "chicks."

In her *Harry Potter* series, J. K. Rowling has a professor deliver a cautionary lecture to his students before they brew up a love potion: "*Amortentia* doesn't really create love, of course. It is impossible to manufacture or imitate love. No, this will simply cause a powerful infatuation or obsession. It is probably the most dangerous and powerful potion in this room." In their potential to co-opt a person's deepest emotions, Rowling suggests, love potions are sinister. They mimic the lack of emotional control that comes with infatuation, but act wholly outside the affected person's free will.

In another book we'd happily explore why love potions and anti-love potions have been such powerful and enduring tropes in fiction. Perhaps it's because they work as a desirable fantasy for anyone who's ever experienced an unrequited crush, or because they capture some of the very real and disruptive effects of passionate love. Or maybe they're just handy plot devices.

Our interest here, though, is in real-life neurotechnologies that act on the brain's lust, attraction, and attachment systems, whether to strengthen a good relationship or help end a bad one. Although we have been referring to these interventions as love drugs and anti-love drugs for convenience, these neurotechnologies are altogether different from magical love potions. For one thing, love drugs (as we have been using that term) are real; for another, they cannot completely override a person's free will, rendering them as pliable as a character in somebody's play or opera.

As we will explore later, a major way to affect love in real life is to manipulate hormone levels. As the science writer Kayt Sukel notes in her book *Dirty Minds: How Our Brains Influence Love, Sex, and Relationships*, hormones have something of an outsized reputation: "From our earliest years," she writes, "we are told that hormones can influence everything, from our boobs (or balls, as the case may be) to our brains to our behaviors. Hormones will flood our system. They will take over and rage out of control." But we are not actually slaves to our hormones, especially once the storm of puberty has subsided. To see this, Sukel says, just consider the common rat:

> Hormones actually control sexual behavior in this species—not mediate or motivate, but control. The female rat ovulation cycle lasts 4 to 5 days. When the female is at her most fertile, the hormone levels rise and her back arches up, exposing her private parts to the world. This reflex is called lordosis. It's a sign to all the boy rats that the girl is ready to go . . . [she] does not have to consider whether she feels up to sex after a long day of running mazes in the lab. There is no worry about emotional readiness or whether she looks fat. Her hormone levels let her know it is time to get busy. So she does. Otherwise she cannot be bothered. It's just that simple.

Not so with humans. According to Kim Wallen, a neuroendocrinologist at Emory University's Yerkes National Primate Research Center, "Hormones are not absolute regulators of behavior. The function of hormones is to shift that balance of behavior in one direction or another. The presence of certain hormones doesn't mean you will exhibit a certain behavior but rather increases the probability that you might."

The same thing is likely to be true of most real-life biochemical interventions into love and relationships, both now and in the future: there are no actual magic potions out there that will instantly transform your emotional life, making you fall out of love in a heartbeat with your spouse of thirty years, or in love, for that matter, with every pizza guy who shows up at your door. As the anthropologist Helen Fisher explains:

> As you grow up, you build a conscious (and unconscious) list of traits that you are looking for in a mate. . . . Drugs can't change [this] mental template. Altering brain chemistry can [influence] your basic feelings. But it can't *direct* those feelings. Mate choice is governed by complex interactions between our myriad experiences, as well as our biology. In short, if someone set you up with Hitler or some other monster, no "slipped pharmaceutical love potion" is going to make you love him.

In other words, the most likely scenario for the foreseeable future, even as neuroscience progresses, will be more or less powerful loadings of the dice—not sorcery.

Unmasking love

Even so, we know that love drugs will be divisive. Some people, we presume, will welcome the advent of drug-assisted romance. "Better relationships through chemistry," they might say (riffing on the old DuPont advertising slogan). Others will find the prospect of biochemical interventions into romantic relationships at the very least unsettling, if not abhorrent. Love, they might think, is not to be tampered with. It's something you're supposed to fall into, spellbound, if you are lucky enough to meet the right person. It isn't meant to be under our control.

We get it. We've been in love. We've experienced the magic. And we understand why pulling back the curtain and talking about hormones and neurotransmitters might spoil the thrill. But we want to push back against this intuition, too. There can be great value, we suggest, in regarding love as something that is—at least to an extent—up to us. Something that requires choice, skill, and determination, not passivity and acquiescence. It is true that we cannot simply wave a magic wand to bring love in or out of existence (nor should we necessarily want to); but we *can* decide whether and how to intervene in the course of love, helping it to last or, where appropriate, expire.

This is not as radical as it sounds. Humans have long meddled with the chemistry between lovers. It's just that our capacity to do this effectively is about to pick up speed. Take couples counseling. If you think this can be an acceptable, even praiseworthy, course of action for at least some romantic partners, you should already be comfortable with the idea that love "takes work." Normally, of course, this work involves making psychological and behavioral changes, not direct changes at the level of the brain. But why not both, as we suggested with the example of Sofia—the woman from the beginning of this book who reached out to us seeking a chemical breakup with her misogynistic husband?

Our thesis is that people—individuals, couples, relationship therapists, scientists, regulatory agents, lawmakers; in a word, society—should seriously consider the prospect of *complementing* psychosocial interventions with interventions into love's biological side. To ignore this latter dimension is to obscure a crucial aspect of the ties that bind, and we ignore it at our peril.

Just think: everyday activities like holding hands, sex, intimate massages, and other sensual activities play a huge role in cementing the romantic connection between couples. In major part, this is because of the social context, subjective experiences, and special meanings associated with those behaviors. But no less significantly, it is because they unleash a flood of neurochemicals that reinforce attachment directly in the brain. These neurochemicals, in fact, work a lot like recreational drugs—think heroin or cocaine—making you feel literally addicted to your partner. So parents who warn their teenagers to be careful about who they shack up with might have a point. Sex-fueled attachment doesn't wait around for evidence of emotional compatibility between lovers before starting to form. Instead, erotic touch, sex, and orgasm can induce hard-to-shake feelings of intense interpersonal connection, very quickly and fairly directly, through the deluge of chemicals they trigger in the brain.

One of these chemicals, oxytocin, will get a full treatment later on, along with its usual co-conspirators, dopamine and vasopressin. The popular media portrays oxytocin as a cuddle drug or a love hormone, but as we'll see, the reality is much more complicated. Nevertheless, oxytocin is crucial for sustaining the attachment bond between mothers and their infants (through breastfeeding, for example), as well as the bond between adult lovers. In fact, the two bonds are closely related, with a lot of anatomical and neurochemical overlap. Basically, if your brain stops producing enough oxytocin or stops processing it in the right way, you can cuddle and hold hands as much as you like, but you won't form a pair bond with your partner.

Oxytocin (and other brain chemicals) can now be produced in the lab. You can even order oxytocin off the internet in some places and spray it directly up your nose and into your brain, though we don't recommend it. The people currently doing this, thankfully, are mainly scientists, who are spraying it not up their own noses but up the noses of volunteers they recruit for carefully controlled studies. Later we'll describe an experiment that looked at the effects of such artificially administered oxytocin on couples engaged in an argument. The upshot is that it seemed to increase positive communication behaviors and reduce stress in this otherwise antagonistic situation. These factors have been shown to play an important role in predicting long-term relationship survival.

But why use a nasal spray when you could have sex instead? Why not let your brain release the chemicals naturally instead of snorting them out of a bottle?

These are good questions. Certainly, taking advantage of your body's natural bond-promoting mechanisms is a sensible thing to do. But it doesn't have to be either/or. The nasal spray could—and should—be used as an adjunct to other interventions, if and when it becomes more widely available. Another issue is that some people really struggle with physical intimacy (because of shyness, shame, past traumas, or other reasons) and may need some chemical assistance to get the ball rolling. Or take this example from the online comments section of a *Jezebel* piece discussing one of our papers:

> I just rounded the corner of my 4th year of marriage, and my husband is in his 3rd year of a medical residency—I see him awake 3x a

> week if I'm lucky. And that awake time usually translates to dinner, conversation, and then sleep. You know what that means? Unless I'm immediately stripping off both of our clothes to have sex I'm not really in the mood for (or he isn't really in the mood for; his job is very stressful!), we maybe get a single release of oxytocin (from sex/touching) a week. That makes you feel really unconnected to a partner, even if you talk . . . and touch as often as you can.

Another commenter had this to say:

> The hard thing about constant relationships . . . is that human moods and vagaries are stronger and less reasonable than we often admit. You can wake up feeling completely different about somebody than you did the night before, or an unrelated situation can make you feel negatively about many things in your life. But that's a feeling, it will change. Marriage, and deep love are more than the changing feelings. A committed relationship is a . . . choice, not something that happens to you and you live happily ever after. If this oxytocin nasal spray thing will help people in committing to their choice and improve [the] quality of relations and quality of life for some couples, then I think it's a wonderful idea. If you don't think it's a wonderful idea, don't use it.

Again, this idea is not entirely new. In certain ways, the chemical transformation of love—especially its sexual side—has been going on for many years. Viagra (sildenafil) has been used since the late 1990s to help men maintain erections. It isn't a love drug exactly, but still, it can work as a romantic aid for those couples who continue to value a particular style of sexual intercourse as they age. And now Addyi (flibanserin), dubbed "the female Viagra" by some in the media, is being controversially touted as a prolibido drug for women.

Drugs with antilibido effects exist as well. Testosterone blockers, for example, are sometimes injected into convicted sex offenders, and this has been going on for decades. Such drugs can powerfully dampen sexual urges, not only in men but also women (and presumably people of other genders). They can also cause impotence, depression, and osteoporosis.

It isn't all about sex, though. The emotional side of love can be affected as well. For example, some drugs used to treat depression have a side effect of sapping people's ability to care about their partners' feelings—a pretty basic ability in any functioning relationship. And there are other drugs that may make it *easier* to care about a partner's feelings, like MDMA, the "empathogen" we mentioned before. Again, all of these drugs exist right now. Still others have yet to be created.

Love or something lesser

Romantic love is about more than biology. And good relationships are about more than popping pills. But interventions into the biological side of love have received much less attention than interventions into its psychosocial side, and we think this needs some evening out. The reason for this is simple. Such biointerventions have the potential to dramatically alter the experience, quality, or even existence of love between individuals, on just about any plausible definition of that term. To illustrate, we'll pick one main feature of love that most people agree is important, and show how it can be partially manipulated through biochemical means.

Consider the view that true love, whatever else it is, is something that requires genuinely caring about (and trying to promote) the other person's well-being as an end in itself. At a minimum, then, you have to be seriously invested in the other person's feelings—not just their transient emotional states, but also their deep desires and preferences, their wishes and dreams, their complex experiences, and all the meanings they carry.

Now imagine that you take a drug that makes it so you *don't* care about your partner's feelings in some or all of those senses, much less their overall well-being. Or perhaps you do care, but only in some abstract, cognitive sense that doesn't correspond to the appropriate motivations or behavior. Suppose you can see that your partner is very upset about something, for example, but their being upset doesn't strike you as all that important (as long as you are taking this drug). You know it should affect you. It isn't that you think they are pretending. You believe they are genuinely upset and have good reason to be. But their anguish just doesn't move you.

Does such an awful-sounding drug really exist? Yes, it does. It's called a selective serotonin reuptake inhibitor, or SSRI, and it's the most commonly used drug to treat depression. SSRIs don't have this effect on everyone. But there are plenty of case reports of people taking SSRIs who experience "diminution in emotional responsiveness" to those around them.

To explore this phenomenon, the psychiatrist Adam Opbroek asked his patients who reported SSRI-induced sexual dysfunction to fill out a number of questionnaires to assess their emotions. Compared to a control group that was not taking SSRIs, these patients reported less ability to cry, experience irritation, dream erotically, or express creativity; to feel surprise, anger, or sexual pleasure; to worry over things or situations; and, importantly for our purposes, to care about the feelings of others. Fully 80 percent of the patients in Opbroek's study described "clinically significant blunting of several emotions."

These results should not be surprising. Part of the point of SSRIs—at least in their role as a treatment for depression—is to "blunt" certain emotions, in particular one's own maladaptive feelings of sadness. But for some patients, the ability to care about other people's feelings seems to be blunted as well. What if one of those other people is your romantic partner? Remember that we are assuming that caring about your partner's feelings is one of the bare-bones necessary ingredients of true love. If your very capacity to do this is sufficiently degraded by an SSRI, over a long-enough period of time, then the drug will by definition change your love for your partner—potentially to the point that it no longer counts as love at all. Change in biology, change in love: proof of principle.

Now consider the loss of libido that often occurs as a side effect of SSRIs. How might that affect your relationship? For many couples, sexual interaction is a very important part of how they relate to one another. Indeed, on some views, the experience and expression of sexual desire is a major part of what makes romantic relationships what they are. In other words, wanting to be physically intimate with someone (more than fleetingly or only under unusual circumstances), at least in the early stages of a relationship, is often seen as the singular thing that carves a more or less distinct dividing line between the sort of love you feel for your romantic partner and the platonic love you feel for your best friend—no matter how strong and sincere your feelings of affection for the latter.

If that sort of view is right, then a drug that removes your desire for sexual activity with your romantic partner is one that will change something essential about the love between you. At the extreme end, you might think that sexual desire is required for romantic love to exist, in which case the drug in this scenario would be directly responsible for quelling love, simply by lowering your libido below some threshold. But there are also other, less-direct ways in which certain drugs can change, if not necessarily eliminate, your love for your partner.

Less-direct pathways

Let's say you completely reject the idea that a pharmaceutical can change love directly, much less make a person fall in or out of love in the first place. Maybe you think that love is too complicated or abstract to be induced, extinguished, or even just buffeted around by mere chemicals. We think you'd be mistaken, for the reasons we've already given, but we'll concede this point for the sake of argument. Let's consider some less-direct pathways, then, to ease you into our perspective.

We think you'll agree that some drugs, at least, might be able to affect your *motivational states*—your willingness to empathize or spend time with your partner, say, or your openness to hearing their perspective. MDMA, for instance, appears to have such effects on some people. Clearly, those things—empathizing, spending time, listening, and so on—make a huge difference to the quality of love between people, to whether it even develops in the first place, and to whether it persists. In our view, if a drug can shape motivations and behavior in ways that make it nontrivially more (or less) likely that love will come about or survive, then we're happy to call it a love drug (or an anti-love drug), even if it doesn't affect love directly.

Alcohol offers a simple illustration. It may be the oldest and most popular love drug around. In addition to making it easier to talk to someone you might otherwise be too nervous to engage with (talking, of course, is a precondition for most relationships), it can also amplify how attractive you find them, how attractive they find you, and how attractive you *think* they find you, which can boost your confidence, at least in the short term. If you've ever had a few drinks in a social situation, you already know this. But there have also been controlled studies confirming the existence of "beer goggles," both in the lab and in actual bars.

For a twist on the usual finding—that alcohol makes others seem more attractive—consider a 2013 study in the *British Journal of Psychology*, cleverly titled "Beauty Is in the Eye of the Beer Holder." This study concluded that people who are drunk (or even just think they're drunk) also think they are more attractive. To show this, the French researcher Laurent Bègue and his collaborators camped out in a Grenoble barroom and handed out lottery tickets to entice patrons to participate. Instead of rating the attractiveness of someone else (the usual measure in these kinds of experiments), the researchers had customers rate *themselves* on how attractive, bright, original, and funny they felt in a given moment.

On the basis of breathalyzer results, Bègue then correlated these ratings with customers' blood alcohol content and found the reported relationship. Basically, the drunker you are—up to a point—the more you think you are sexy and awesome. And the more sexy and awesome you think you are, the more likely you are to introduce yourself to the cute-looking person across the room.

Does this mean the person will return your interest or find you as fascinating and attractive as you currently think you are? Of course not. Who knows how things will play out. Maybe you two have nothing in common and this becomes obvious the second you're sober. Maybe at closing time, when they bring the lights up, you'll both recoil, and stumble separately home.

Or maybe not! Maybe you do hit it off. Maybe you end up arranging a second date. And then a third. Maybe that initial introduction—that first, smiling, caution-to-the-wind hello that wouldn't have happened without some chemical courage—becomes the "How Your Grandparents Met" story that you tell to your grandkids fifty years later. Countless relationships, loving and otherwise, have started out precisely this way.

We would like to draw a couple of lessons from this example, simplistic as it is. First, as we said, a drug doesn't have to cause love directly to make it more likely to come about (or last). The same thing goes for anti-love drugs and the prevention of love. This book deals with nudges and probabilities, not chemical puppetry and emotional determinism.

Second, just because a drug plays a role in bringing about or maintaining love doesn't mean the love itself is inauthentic. In the bar example, alcohol inspired the conversation, which is a crucial threshold you have to cross. Once the drug wears off, though, you're on your own.

Third, context matters. To stick with the alcohol example, drinking alone in your room is not going to make you fall in love with anyone—and it probably won't make anyone fall in love with you. For a love drug to be effective, your mind-set, the setting, the other people involved, and a whole lot else have to coincide and interact in the right way.

More-direct pathways

What about more-direct chemical pathways to love? Earlier, we mentioned the hormone oxytocin. This hormone is released by the brain in mothers during childbirth and through breastfeeding, and it is largely responsible for the development and persistence of the mother-infant pair bond—a prototypical instance of love on most accounts. This same hormone is released through intimate touch, sex, and orgasm; it is similarly responsible for the attachment that forms, insofar as it does, between adult romantic partners. If you stop having sex with your partner, then—due to the libido-killing effects of an SSRI, let's say—you will also stop releasing oxytocin in one of its most potent contexts, which in turn could negatively affect the attachment bond between you. Again, depending on which core "ingredients" you think should be necessary for love, this change in biology—like the others we've mentioned—could amount to a change in love.

What about love-*enhancing* drugs? Just run these alterations in reverse. Oxytocin, as we mentioned, can now be administered artificially through a simple nasal spray; and testosterone, which can be boosted in various ways, has long been known to increase libido. Again, neither of these would be sufficient for promoting love in every case. But given the right combination of other mental, biological, and social factors, they could. In other words, there is reason to believe that existing biotechnologies are already capable of altering love, whether positively or negatively, through a variety of more- or less-direct routes. Scientists should study these effects, we suggest, by zooming out from individual-level concerns and looking at the consequences of therapeutic drug use for relationships in different cultural contexts.

This research is not a luxury, and it shouldn't be put off until tomorrow. A huge amount of data suggests, and common sense confirms, that the quality of people's close relationships is one of the most powerful predictors of good health and subjective well-being. Close relationships also show up in most "objective list" theories of well-being, which are the theories that say that certain things are good for you whether you agree or not. And some philosophers argue that loving relationships are intrinsically good, or ends in themselves. If scientific and intellectual resources should be directed toward the promotion of human flourishing—as presumably they should—then it ought to be a priority to study the factors, both psychosocial and biological, that influence relationship quality.

If, for example, it turns out that commonly used pharmaceuticals risk degrading relationship quality for many couples, then we should reconsider when and how they are prescribed. Likewise, if other drugs, or the same drugs under different conditions, can improve relationship quality, then their therapeutic potential should be looked into as well. So what are the *interpersonal* side effects of commonly used prescription medications?

Main effects and side effects

Many drug-based treatments have side effects. These are commonly understood to be unintended, and usually unwanted, physical or mental outcomes that affect the person taking the drug. Some of these potential side effects are known to the drug manufacturers; others can occur without having been studied scientifically or even documented in a systematic way. It has become obvious that many widely prescribed medications affect our thoughts and emotions in ways that have not been adequately researched. Analyzing how these drugs alter our psychology is therefore clearly important—but not just at the individual level. We also need to study how current drugs are affecting our close relationships.

There is a small amount of research on this question. Some scientists now speculate that widely used pharmaceuticals may be having dramatic effects on our relationships, including romantic ones. This is because the chemical properties of some of these drugs are known to affect brain mechanisms that have already been tied, in other research, to things like attracting and choosing romantic partners, subjective feelings of being in love, and common behavioral expressions of those feelings.

A key example is SSRIs, as we've discussed. In addition to being a treatment for depression, they are also often prescribed for anxiety disorders. Their name comes from the fact that they block the "reuptake" of serotonin in the presynaptic nerve terminal, which increases the activity of this chemical at the synapse (the little space between nerve cells in the brain). Between 2005 and 2008, about 11 percent of Americans age twelve and up were on some form of antidepressant, with SSRIs being the most commonly prescribed form. While these drugs can be helpful for many people in treating depressive symptoms, allowing them to function more effectively—including in their relationships—they can also have negative effects on love that are not well understood.

One possible effect is linked to the development of impotence (due to SSRI-induced changes in testosterone levels), which can lead to "performance anxiety." This in turn can make some people more likely to avoid intimate situations altogether. Consider this account of a twenty-six-year-old man who had panic attacks that required high doses of an SSRI:

> He soon experienced diminished libido and impotence. A handsome, personable, intelligent man, he was readily sought after by women. However, he ended several relationships because he was too embarrassed about his inability to perform sexually. Although he tried several other medications, he was able to control his panic disorder only with high doses of serotonin enhancers. Eventually he retreated into a social life in which he avoided serious dating. When last evaluated he still confined himself to non-sexual relationships with women.

In addition to indirect effects on love like this one, increased levels of serotonin caused by SSRIs can directly suppress activity in pathways for dopamine and norepinephrine, neurotransmitters that are involved in the production of subjectively reportable feelings of romantic love. The following comes from a letter published in the *New York Times*:

> After two bouts of depression in 20 years, my therapist recommended I stay on serotonin-enhancing antidepressants indefinitely. As appreciative as I was to have regained my health, I found that my usual enthusiasm for life was replaced with blandness. My romantic

> feelings for my wife declined drastically. With the approval of my therapist, I gradually discontinued my medication. My enthusiasm returned and our romance is now as strong as ever. I am prepared to deal with another bout of depression if need be, but in my case the long-term side effects of antidepressants render them off limits.

Not everyone reports similar experiences. Consider this very different profile of a couple in *Elle* magazine:

> Before they got on antidepressants, Susan's tendency to rail at length (about whatever happened to be irking her) exacerbated Will's "extremely self-critical" tendencies: "Whereas in her depression she'd tend to lash out, in mine I'd tend to sink inward," he says. . . . "We were heading down a bad path." Now, though, they agree their marriage is much better balanced. Susan's rough edges have "softened," as Will puts it, and with this—plus the boost medication has given his own confidence—he's become more forthcoming: They're able to work together to solve problems. "We really are each other's best partners," Will says. "To call us soul mates I think would be accurate."

What these contrasting cases show is that there are many different mechanisms through which biochemical substances can exert their effects. Accordingly, SSRIs (or other drugs) can put some instances of love in jeopardy, while in other cases they might serve as love enhancers. Of course, the precise connections between what goes on in the brain as the result of any particular biochemical intervention, and people's subjective experiences of being in love, are complex and currently not well understood. These connections likely differ from person to person and couple to couple. And different doses, timings, and contexts only add to the confusion.

As a result, the full range of effects of SSRIs (and other widely used drugs) on individuals and relationships is mostly unknown. All that scientists can say at this point is that the flexible nature of the brain mechanisms behind sexual reproduction and their complicated interactions with certain drugs give us reason to be concerned. According to Helen Fisher, "any medication that changes the chemical checks and balances is likely to alter an individual's courting, mating, and parenting tactics," with potentially major implications for relationships.

Hormonal birth control

One such medication is "the pill." According to the U.S. Centers for Disease Control and Prevention, between 2006 and 2010 62 percent of American women of reproductive age were using contraception. The most common form of contraception was hormonal birth control, used by 10.6 million women. Hormonal contraception refers to birth control methods that act on the endocrine system. Almost all of these consist of steroid hormones, and there are two main types: combined methods, which contain both estrogen and progestin, and progestin-only methods. Combined methods work by suppressing ovulation and thickening cervical mucus, while progestin-only methods reduce the frequency of ovulation.

As women who use birth control know, common side effects of progestin-only methods are not trivial. They include menstrual cycle disruption, headaches, weight gain, and breast tenderness. Some forms of hormonal birth control can also cause nausea and mood swings and can affect libido in some women as well. While the libido-altering effects could have indirect consequences for relationships, as we've seen, recent research points to a more direct connection between hormonal birth control and intimate relationship preferences in women.

One common theory goes like this: hormonal variation over the menstrual cycle changes (heterosexual) women's preferences for outward signs of men's genetic or parental quality. These may include signals communicated unconsciously via pheromones. Since hormonal contraceptives suppress this variation, different mate preferences might develop than would have otherwise. For example, women who are using oral contraception in the early stages of dating could end up choosing a less sexually compatible partner than they would have if they were not on birth control.

The psychologist S. Craig Roberts and his colleagues have tested this idea. Their main strategy was to look for differences in relationship quality and eventual breakups between two groups of women: those who either were, or were not, using oral contraception when they initially got together with the man who later fathered their first child. Their study found that women who were on birth control "scored lower on measures of sexual satisfaction and partner attraction, experienced increasing sexual dissatisfaction during the relationship, and were more likely to be the one to initiate an eventual separation if it occurred."

Curiously, "the same women were more satisfied with their partner's paternal provision, and thus had longer relationships and were less likely to separate." So the interpersonal effects of birth control might not all be bad. Either way, the authors concluded that "widespread use of hormonal contraception may contribute to relationship outcome, with implications for human reproductive behavior, family cohesion and quality of life." Just as with SSRIs, however, these implications are only barely beginning to be studied.

Shifting norms

These findings, however tentative, suggest a need for a shift in research norms. Take depression. In her article "The Couple Who Medicates Together," journalist Louisa Kamps notes that many standard measures of depression focus on the individual's mood symptoms, like sadness, emotional withdrawal, and lack of appetite, "rather than on how respondents are getting along with spouses, co-workers, and kids."

At the same time, negative and potentially harmful emotions like (maladaptive) anger are "common in depression—and spouses tend to bear the brunt." Kamps speculates that scientists' failure to appreciate this interpersonal dimension could be one reason why antidepressants sometimes get a bad rap: "The data don't capture their ability to reduce volatility [in relationships] because researchers don't ask about it in the first place."

Our suggestion here is simply that they should. And not just questions about volatility or anger, but about all the various factors that go into relationship quality. This could happen in a number of ways. Researchers could add to the protocols they are already using to test drug outcomes, by building in measures of relationship satisfaction. Funding agencies could emphasize the need to study the social aspects of drug use when making decisions about doling out grants. Pharmaceutical companies could be taken to task for advertising drugs without first evaluating their interpersonal implications. Doctors could include questions about positive and negative effects on relationships when checking up on their patients' progress on medications.

However it's done, interpersonal health and well-being should be front and center in studies on the effects of pharmaceutical interventions. It isn't enough to study individuals and their symptoms. Each of us is embedded in at least some kind of social context, and if certain drugs are going to continue to be prescribed for therapeutic purposes, we have to understand their impact on our relationships.

We have looked at some of this impact, based mostly on side effects. But what about the intentional use of a drug to influence a relationship? Next, we introduce you to a couple with very specific romantic issues that they are working through together. Could love drugs be an option for them?

CHAPTER 5

GOOD-ENOUGH MARRIAGES

WE'VE SEEN THAT LOVE-ALTERING DRUGS are already here, partly in the form of understudied side effects of widely used prescription medications. More potent love drugs are on the horizon. We don't want to be caught flat-footed: equipped with the power to biochemically intervene in relationships, but not knowing if, when, or how to intervene. So one of the first things to get clear about is this: how do you decide whether a given relationship is worth preserving—with or without the use of a drug—or whether it would be better, all things considered, for it to end?

If love drugs do become more widely available, this could alter the course of such decisions, because the drugs might change the feasibility of pursuing either option. In other words, they would become part of "all things considered." So the time is ripe to start thinking through likely scenarios, to help us get a grip on when love drugs might or might not be a good idea. If it turns out that we can't think of a situation where love drugs would be at all appropriate, we can save ourselves the trouble.

We'll start with a particular type of relationship that we think might be a promising candidate for drug-assisted couples counseling in the near future. Along the way, we'll share some ethical tools that individuals can use to decide whether to stay in a long-term relationship or leave it, given the sorts of dilemmas that many modern couples face.

Stella and Mario

Meet Stella and Mario. She is thirty-eight, he is thirty-nine. They have two young children, a boy and a girl. They married thirteen years ago, after several years of dating minus a few months here and there for soul-searching and time alone. Taking a break back then wasn't the end of the world. They weren't tied down. They even dated other people, casually. But they kept coming back to each other; they couldn't help it. Friends and family started to say that it was inevitable. Everyone knew it was coming. Finally, they tied the knot.

In the early years of their marriage, life was good. It was comfortable and thrilling all at once. They seldom fought. They had mutual friends. There was time for nights out when they were up for it, and nights in when that's what they preferred. They laughed. A lot. They respected each other's space, but never held each other at arm's length. And best of all, they could get on with their busy professional careers—she as a lawyer, he as a physical therapist—secure in the knowledge of each other's support. At parties, they introduced each other as "my partner" and meant it without self-consciousness or irony. Theirs was a loving, fulfilling relationship by any measure.

After they had children, things changed. To be sure, the high of new life—the miracle of creating a human being, and then beings, together—was intense. It brought them much closer at first. But then fatigue came, and became a constant. Priorities changed. Of course, the kids were wonderful. They were intelligent, playful, and kind. They turned in their homework on time; they were curious about the world; they made their parents smile, mostly. There weren't any significant illnesses or other calamities. Still, their parents' relationship to each other took a turn for the worse.

For her part, Stella is riddled with guilt. Her full-time career is turning motherhood into a millstone, propped up by increasingly infrequent fits of "quality time" spent with the kids between work and sleep. Sometimes it's forced. Always it's in the teeth of exhaustion. Time with Mario is not even on her radar. And for his part, fatherhood has become strained as well. He takes the kids on trips sometimes, and that seems to help. But his feelings of love—romantic, erotic, passionate love—for his partner, so strong in the beginning, have faded. Stella is starting to feel the same way.

Each of them has a relationship now as parents to their children—not with each other. They have stopped having sex or enjoying it when they do. It has become a mechanical act they feel at times obliged to engage in. Is this what life is now? Nights spent snoring in bed with backs turned against each other? Bickering in the morning? Fretting over what to do with the kids on weekends? The early bliss of marriage is starting to feel more like a cage.

They have tried counseling. It didn't seem to work. They went on a couple's retreat. Their feelings for each other did not rekindle. They even opened up the relationship to other partners but decided this was not for them. In fact, they tried everything they could think of, and still no flame.

So why not get a divorce?

They've thought about it. They've talked about it. On their worst days they've sworn to themselves, and sometimes to each other, that they'd *do* it, too—only to take it back in a stream of tears. There are the kids to think about. And anyway, it's not like they hate each other. Their values are basically in line. They have a lot in common: shared routines, shared finances, a shared sense of humor, even, when the rare opportunity to appreciate that fact still arises.

And their shared history is not a minor consideration. Their lives, their memories, their identities—for better or worse—are intertwined. You pull the threads apart, and you unravel. But there is no getting around the fact that the color has simply drained from their relationship. It's gray now. All the time. And there is no apparent escape.

The dilemma of gray relationships

What you just read is based on a true story, told to us as we were writing this book. Or rather, many true stories. Maybe you have felt this way yourself; it's a common predicament. And if you have felt this way, or feel this way right now, you know how superficially easy it is to say, "Just get a divorce—move on—find a relationship that makes you happy." Or maybe you've heard, "The kids will be fine; you have to do what's best for you." And sure, that might be right. There *are* couples in "gray" relationships who find that despite all their reservations, separating and moving to different houses, maybe splitting up the kids, and finally seeing what Tinder is all about is the best way forward.

Others can't bring themselves to take this step. It feels wrong, deep down; and maybe, for them, it is. This doesn't necessarily mean that they are weak or insufficiently concerned with their own well-being. Sometimes it is an honest reflection of the complexity of life and relationships: you made a commitment, you love this person, you want to find some semblance of joy together—but you've run out of places to look.

In her book *Marriage Confidential: Love in the Post-Romantic Age*, historian Pamela Haag writes that of the more than one million divorces that occur annually in the United States, the majority come from a population you might not initially expect. It isn't high-distress, high-conflict marriages marked by "abuse, violence, addictions, fistfights, chronic arguments, projectile shoes and dishes" that keep the divorce court doors swinging. Instead, she says, it's relatively low-conflict marriages. Ones that are fairly stable, more or less functional, occasionally pleasant, but, when it comes right down to it, decidedly less than happy. Something like what we described with Stella and Mario.

In these marriages the choice is not as simple as being trapped and miserable, on the one hand, versus staging a dramatic prison break that leads to glorious freedom, on the other. Instead, it's between persisting in a kind of low-grade melancholy for an indefinite period of time, or making a serious, possibly wrenching change. If you are a spouse with "vague feelings of discontent" (as Haag puts it) or "the problem that has no name" (as Betty Friedan put it), or if you are married to someone who has such feelings and you worry about being inadequate for them or holding them back, then you know the kind of marriage we have in mind.

There is no magic formula here. No single solution to the question of whether, or when, a couple should throw in the towel when faced with marital dissatisfaction. Some relationships should definitely end, particularly if they involve violence or other forms of abuse. Other relationships have enough going for them, despite the difficulties, that the couple could reasonably decide it's worth trying to make things work. A lot comes down to the specifics, to alternative courses of action that may be open or that have already been tried—and to the couple's values. We won't be giving any easy answers. But between this chapter and the next, we will highlight some of the more promising ethical strategies you might use to try to resolve such dilemmas.

Autonomy as an ethical tool

One strategy is to focus on autonomy. This principle says that all else being equal, mature adults should be free to choose what they consider to be best for them, even if others think their choice is foolish, or not in their best interests, or simply not what they would do. So if you believe that ending your relationship is the right choice—no matter what your partner, friends, family, counselor, priest, rabbi, or anyone else may suggest—then you should be entitled to end it.

We agree with this view, at least broadly speaking, and we'll explain why in a moment. Certainly, we will not be rehashing old debates about the moral permissibility of divorce or other means of ending a relationship. If someone beats you up, manipulates you, or makes your life a living hell, it's imperative that you be free to leave. Period.

But what if things are not as bad as that? What if you are like Stella and Mario, a couple who promised to stay together, who aren't outright abusing each other or tearing each other apart, but who also can't seem to make each other happy?

These are difficult situations. If promises (such as wedding vows) mean anything, they *should* be hard to break. In fact, the whole tradition of promising, on which so much rests in society, starts to crack at the foundations if we begin to take it lightly. If saying "I do" at the altar comes to be seen as having your fingers crossed behind your back, it loses its power to usher in a new, shared reality and to secure the particular benefits that go along with a solid commitment. (Of course, there are benefits that go along with freedom and independence, too, so it's a matter of trade-offs, like so much in life.)

This fact about promising has at least two implications: (1) you should be very thoughtful about the promises you make, especially if they have lifelong fulfillment conditions; and (2) if you go ahead and make such a promise, assuming you weren't coerced, you have an obligation to honor it, within reason.

This doesn't mean you could never be justified in leaving a "merely unhappy" marriage (or other committed relationship) if that is what you determined was best. The balance of reasons in favor of this choice might be less decisive than in cases of, say, intimate partner violence or other examples of clear dysfunction. But in addition to autonomy, the sheer pursuit of happiness is a fundamental value. And the scale for weighing up that pursuit against other values (like the value of keeping a promise) is ultimately in each of our own hands.

But why exactly is autonomy a value at all? Now we'll explain our support for this principle, so it isn't just taken for granted. As we see it, the value of autonomy is rooted in a simple and intuitive idea, namely that individual adult human beings—as opposed to the government, or religious leaders, or the larger community in which someone has been raised—are usually best positioned to know what is most likely to promote their flourishing.

This is not a fail-safe rule, and we should always be open to the insights of others, whose very distance from our situation can give them a useful perspective. But as a foundational value, in the world we live in, autonomy deserves to be front and center. As a point of contrast, top-down norms and value systems, especially those with a one-size-fits-all set of rules, have a long history of being short-sighted, narrow-minded, and downright oppressive, particularly toward nonconformers, minorities, and other vulnerable people. That's not a very promising pedigree.

Yet there are life values other than autonomy. Ethical approaches based on community, for example, recognize our shared commitments and interdependence. These approaches have a lot to contribute. In fact, there may be some societies, or at least periods of history, where even very strict community-based norms prescribing absolute conformity might be preferable to the ethics of autonomy (insofar as the two came into conflict). In societies where the idea of the individual as the basic unit of moral analysis is not well established, autonomy might not trump all other considerations.

Let us use an extreme example to illustrate this point. Suppose you live in a closed society, where members have to count on each other absolutely to ward off threats from hostile outsiders; where lock-step social coordination is necessary for sheer survival; where there is little or no access to other points of view that might shake your confidence in the way of life with which you are familiar. This is quite plausibly the situation in some reaches of our planet, where it might be the case that strictly enforced norms and values limiting individual choice are the best way, practically speaking, to ensure that community members live happy and fulfilling lives.

But this does not describe most modern societies, including the ones in which we assume our readers live. Most of us *don't* live in isolated communities with limited access to other perspectives. We mix and mingle and travel and have a world of information at our fingertips. And we typically don't need strict conformity to defend ourselves from enemies. Apart from the discipline required in military settings to maintain an actual fighting force, what liberal societies need to defend themselves from external threats is creative thinking, open-mindedness, innovation, and cooperation.

Because we are constantly exposed to different visions of the good life, propping up just one of them with rigid enforcement mechanisms is bound to fail. Given our rich, pluralistic world, where many visions of a good life can be defended in terms of shared, fundamental values, it is in everyone's interest that people have a lot of flexibility and freedom to match their life trajectories to a vision that works for them.

If that means getting a divorce, or separating from your partner, to pursue your own idea of happiness, well, the principle of autonomy has your back.

Unhappily married, with children

Sometimes there is more than just a partner on the line when a relationship begins to crumble. What happens when children are involved? As feminist philosophers have long argued, ethics is not just about me, me, me. Instead, we are all dependent on others, to a greater or lesser extent, at different phases of our lives and in different situations. Our ability to be autonomous at all presumes that we have been cared for in a social environment and provided with opportunities to develop our capacities.

As a consequence, couples with children, especially young children, face a different sort of moral decision when considering a breakup or divorce than couples who don't have kids, no matter how fashionable it is to say "the kids will be fine" and leave it at that. As with most things in life, the truth is a bit more complicated.

To put our cards on the table, we think the strongest contender for a relationship in which love-enhancing drugs would be in principle justified is a case of this kind. A case of committed partners who are in a gray (rather than violent or abusive) relationship, who want to make their relationship better, and who have youngsters in tow who depend on them for care and support.

Paul Amato is a social scientist at Pennsylvania State University who's been studying marriage—and divorce—for decades. The popular narrative says that divorce is not all that threatening to children's welfare and might even make them happier in most cases, since their parents are no longer at each other's throats. But things aren't that straightforward. Amato has concluded from his extensive research that children's well-being after divorce depends in large part on the kind or level of stress and antagonism between the parents before their decision to separate. In a nutshell, "when discord is high, divorce appears to benefit children, but when discord is low, divorce appears to harm children." Even that is far too simple, as we'll see, but the point is that divorce is not a monolithic experience for parents or their children. The devil is in the details.

Making responsible decisions in relationships means dealing with those details. Black-and-white ethical thinking is simply not the way to go. For example: "Divorce is always good for children. They are better off with two happy single parents than a sad couple dealing with conflict." Or: "Divorce is always bad. It is always harmful to children (or against the will of God)." Things are rarely black and white in the real world. They are, like the relationship between Stella and Mario, at least a little bit gray.

For some children, parental divorce is beneficial; for others, it is truly harmful. This second group is sometimes swept under the rug, usually in the rush to defend the right of adults, especially women, to leave bad relationships. Of course, women are usually expected to do the lion's share of childcare, typically without compensation or even decent social assistance. This means that "do it for the children"-type arguments tend to have asymmetrical implications for mothers versus fathers, assuming a heterosexual couple. This asymmetry should not be discounted (nor of course should the experiences of gay and lesbian couples, whose decisions surrounding divorce have been much less studied). At the same time, the nuances of children's welfare in response to different parental choices should not be waved away.

Since the right to leave a marriage is so important, the thinking often goes, it must be on balance good for everyone when parents actually do get divorced. But this doesn't follow. Something might be a fundamental right or even just a norm that is worth defending at great cost, while still sometimes resulting in negative outcomes for some of the people involved. We need to be alert to the possibility that divorce may harm some children.

For one thing, the very process of getting divorced (and its aftermath) can lead to such despair and depression in some people that it stops them from parenting effectively, sometimes for years. People can get caught up in reshaping their own lives and careers, or even just coping. Sometimes they become oblivious—or unable to respond—to their children's needs. When these children grow up and are themselves adults, they may face enormous insecurity in their own relationships. Many of them fear that the bond they share with a partner will one day fail, just as it did with their parents.

Readers whose parents separated when they were children might recognize these fears in themselves. But others will not. For some, life took a clear turn for the better when their parents, who had been constantly fighting, finally broke the cycle of hostility by splitting up. Adjustments like having to shuttle back and forth on weekends might have been disruptive, or might have been a fun way to mix things up, or both, or neither; but for some readers, life after their parents separated felt like heaven compared to before. As Amato writes:

> For some children, divorce is a relief from an aversive, dysfunctional home environment; however, for other children, divorce is the beginning of a downward spiral that can last a lifetime. Correspondingly, some marriages cannot be salvaged, and ending them expeditiously is the best outcome for all concerned. But other [unhappy] marriages—marriages that are "good enough" from a child's perspective—*can* be salvaged, and finding ways to help these parents stay together promotes children's best interests.

These are the marriages we have been calling gray, or as we might also say, "good enough." A good-enough marriage might *not* be good enough for some individuals intent on personal happiness, and we have suggested they have every right to leave. But such marriages often *are* good enough to make a positive difference in the lives of children when they're involved, and this is one good reason to try to make things work.

In cases like this, we suggest, love drugs could change the calculus. Not all by themselves, of course, but in the context of a lot of hard work, plus therapy, whose healing effects such drugs might one day enhance. Then, instead of sacrificing your own happiness and remaining in a gray relationship for the sake of your kids, you might be able to have your cake and eat it too: do what you believe is best for your children *and* bring love and happiness back into your marriage. Or at least give love and happiness a better shot.

CHAPTER 6

ECSTASY AS THERAPY

AUTUMN (NOT HER REAL NAME), a woman with autism, first took Ecstasy in 2007. She was at a party in Sarasota, at the New College of Florida. Using a testing kit she'd bought from an organization named Erowid—a harm-reduction nonprofit that educates people about the risks of psychoactive substances—she confirmed, as best she could, her drug was pure. She swallowed it and waited to see what happened. As she later told a journalist, the disjointed, difficult-to-process jumble of thoughts she'd grown used to, which she attributed to her autism and had experienced her entire life, began to fall into a more coherent pattern. Later, when she heard about a study on "autism, feelings, and MDMA," she signed up.

"The usual barriers that exist within my headspace—my usual circuitry—becomes more fluid and I can find the contours of the issue or the thing I'm usually unable to obtain." That's what she said about MDMA to the researchers who conducted the study. Autumn said she was "buoyed" by her personal experience and decided to take MDMA with her romantic partner to "tackle some heavy issues." She wanted to share her feelings about the devastating sexual abuse she'd suffered at the hands of her ex-husband, which she had previously been reluctant to open up about. The trauma from that abuse hasn't gone away. But Autumn claims that with the help of MDMA, she and her new partner have been able to at least broach that difficult history, "to some really interesting results."

In the 1980s, before it was made illegal, the psychoactive substance 3,4-methylenedioxymethamphetamine or MDMA—popularly known as Ecstasy due to the feelings of euphoria it can induce—was being used as an aid in couples therapy by professional counselors. Writing in the *Journal of Psychoactive Drugs* in 1998, psychiatrists George Greer and Requa Tolbert described a method of conducting MDMA-enhanced therapeutic sessions based on their experience with roughly eighty clients between 1980 and 1985. After careful prescreening and obtaining informed consent, Greer and Tolbert met with the clients in their homes, believing that a more personal setting would be best for facilitating trust and comfort.

"We never recommended an MDMA session to anyone seeking to be a passive participant who would be 'cured' of [a] psychological problem," they wrote. "We believed that the person treated or cured themselves, with the assistance of MDMA and their relationship to us."

Depending on the clients' preference, they would start the session with meditation or prayer. They then administered a pure, controlled dose of 75 to 150 mg, adjusting for the client's sex or body mass, with a 50 mg booster if requested later on. Clients wore eyeshades to shut out visual distractions and reduce the risk of overstimulation. While waiting for the drug to take effect, they listened to classical music, usually through a pair of headphones (Mahler and Beethoven were among the more popular choices). Then, when they felt ready, clients spoke with their romantic partner.

Often, they would speak for hours.

Not everyone had a major breakthrough. But some did. Three years after her treatment, one client, the thirty-something daughter of Holocaust survivors, wrote that she had previously been prone to anxiety attacks and intrusive thoughts about concentration camps, triggered by her parents' horrific recollections. With the treatment she no longer got caught up in the troubling thoughts and moods that had always harmed her relationships. She now understood better how her emotions worked, she said, having gained some control over her feelings.

Through their own research and that of other pioneers, Greer and Tolbert came to argue that MDMA—administered in the right way, with the careful oversight of a trained professional—could help some individuals achieve self-understanding by decreasing irrational fear responses to perceived emotional threats. This is exactly the theory behind the MDMA-for-PTSD trials we covered earlier. Only it shows that fear and emotional repression can be a problem for relationships even when they don't rise to the level of a clinical diagnosis—and that MDMA-assisted psychotherapy may be helpful in those cases too. According to Greer and Tolbert, about 90 percent of their clients benefited from MDMA-assisted therapy, some reporting that they felt more love toward their partners and were better able to move beyond past pains and pointless grudges.

The effects of MDMA wear off after a few hours. Especially at lower doses, the drug doesn't seem to significantly distort perception, thinking, or memory. According to Greer and Tolbert, their clients would often consolidate lessons learned during the therapy sessions and then apply them to their everyday lives. Romantic couples who experienced a session together "frequently reported basing their relationships much more on love and trust than on fear and suspicion." But these results weren't simply caused by MDMA. They were caused by the clients "making decisions based on what they learned during their MDMA sessions, and [by] remembering and applying those decisions for as long as they were able and willing after the session was over."

This is an important point. If drugs like MDMA are ever to feature in a responsible plan for relationship healing or enhancement—assuming this becomes legal—they should be used in a facilitating or adjunctive role. Such drugs should never be taken in a vacuum, alone or with unprepared others, without the right mental or emotional groundwork, or with the expectation that they will induce improvements all on their own. They won't.

This caveat is classic. In his 1954 essay "The Doors of Perception," Aldous Huxley wrote about his experience taking mescaline, a cactus-derived drug most commonly used today by members of the Native American Church. Mescaline and MDMA have similar effects and they overlap in how they work on the brain, so Huxley's experience can ground a useful analogy. According to Huxley:

> The mescaline experience is what Catholic theologians call "a gratuitous grace," not necessary to salvation but potentially helpful and to be accepted thankfully, if made available. To be shaken out of the ruts of ordinary perception, to be shown for a few timeless hours the outer and the inner world, not as they appear to an animal obsessed with survival or to a human being obsessed with words and notions, but as they are apprehended, directly and unconditionally, by Mind at Large—this is an experience of inestimable value.

Notice that the drug was "helpful" but "not necessary" in bringing about the experience that Huxley found so valuable. This brings us to a related point that applies to other psychoactive drugs. Take psilocybin, the active ingredient in magic mushrooms. According to William Richards, a Johns Hopkins psychologist and leading expert on psilocybin, the drug cannot be taken, at least not responsibly, as a stand-alone medication, independent of mental and emotional preparation and careful consideration of the therapeutic setting.

"One cannot take psilocybin," he says, "as a pill to cure one's alienation, neurosis, addiction, or fear of death in the same way one takes aspirin to banish a headache." Instead, psilocybin—along with the other classic psychedelics and MDMA—provides "an opportunity to explore a range of non-ordinary states. It unlocks a door; how far one ventures through the doorway and what awaits one . . . largely is dependent on non-drug variables."

More than a century ago, a more famous William offered a similar perspective on the altered states of consciousness brought on by his use of nitrous oxide. In his 1902 masterpiece, *The Varieties of Religious Experience*, the great American philosopher and psychologist

William James wrote that such drug-induced, subjective changes "may determine attitudes though they cannot furnish formulas, and open a region though they fail to give a map."

What do these observations suggest for the potential future use of MDMA in couples therapy? One thing they suggest is that the drug will not be doing all the work. Nor will it cause romantic partners to mindlessly transform into starry-eyed zombies, incapable of engaging rationally with their feelings or with the real-life dynamics of their relationships. Rather, couples are still present in the (heightened) experience, taking in the environment, forming memories, accessing emotions, drawing connections, and so forth.

After a few hours the drug wears off and the users return to their normal conscious states. At that point they can reflect on how they felt and whatever thoughts they had during the experience, and decide if there were any valid insights. In other words, they can (and should) integrate their MDMA experience with their more sober experiences of the world and test whatever new ideas they might have against that more durable reality.

This doesn't mean couples will decide to stay together. Maybe they will realize that it would be better to separate and say their goodbyes from a place of love. In the course of writing this book, we talked to individuals who had tried MDMA in the late 1970s before it was prohibited and found that it helped them put their romantic relationship into perspective. Think of a magnifying glass. In one sense, it transforms what you see—and you wouldn't want to peer through it all the time—but it also lets you take in more information than is available to your normal vision. It lets you see things at a different scale and in a different way; things that would otherwise be hidden out of sight.

Some psychoactive drugs can have a similar effect. One woman told us that she experienced, under the influence of MDMA, what love could really feel like directly—including love for herself. She realized that she didn't feel love like that in her relationship, and that something about this was not right. She had let herself get stuck with a partner who didn't value her; she had made a million excuses for why they should stay together, and she didn't have the courage to break things off. But from her experience with MDMA, she learned that she was at least capable of feeling a rich and unfettered love (or something with the same phenomenology) and she became determined to replicate that feeling in real life. She eventually ended the relationship and got to work creating more sustainable conditions for love—love like what she experienced on MDMA—to emerge naturally with another partner.

This is just an anecdote, and as we have said, anecdotes are not enough. But they can be useful for raising questions and formulating hypotheses that might be tested in a more systematic way. From this case it appears that the use of MDMA, at least for some people under some conditions, can foster genuine insights into their lives, with ramifications for their romantic relationship. Similarly, reading the accounts by Greer and Tolbert of their work with clients in the 1980s certainly suggests that drug-enhanced couples therapy could do a great deal of good.

But we cannot go with mere suggestions. What we need is a cautious research program into the possible uses of biological interventions to benefit romantic relationships. Drug-based interventions inspired by this line of research, once they've been carefully vetted through clinical testing, might then be used to supplement more traditional therapies tailored to the needs of individual couples.

Going underground

In 1985 the U.S. government declared an "emergency" ban on MDMA, putting it on the list of Schedule I drugs, which is the official designation for substances with "no currently accepted medical use and a high potential for abuse." The perceived emergency had to do with teenagers taking MDMA at raves and dancing all night in packed warehouses or auditoriums. And in that context there was reason to be seriously concerned. Some young people were dying from not drinking enough water or from combining MDMA with alcohol and other drugs. There are other risks, too, which we'll explore below. But the actual basis for the federal ban—recommended at the time by the Drug Enforcement Agency (DEA)—has been scrutinized in recent years by historians and other researchers, who argue it was uninformed and premature. What follows is an account of a telling episode in the government's decision-making process.

On July 27, 1984, the DEA put out its initial recommendation that MDMA be listed as a Schedule I substance. Exactly one month later, a concerned group of doctors, therapists, and scientists asked for a hearing from the DEA to discuss the proposed scheduling of MDMA. Their goal was to have the drug taken off the prohibited list or at least reclassified so that the research they had been doing, both lab-based and therapeutic (working with clients), could continue. According to one historian, "This request astonished the DEA." The DEA's top chemist, Frank Sapienza, said, "We had no idea psychiatrists were using it."

In part to recover its bearings, the DEA did grant the hearings, which were held in 1985. After listening to witnesses from both the DEA and the concerned group of doctors and scientists, the judge overseeing the hearings concluded that MDMA did not have to be put on Schedule I. He argued that MDMA "had proven its therapeutic potential, can be handled without grave dangers in medically supervised settings, has no high potential for abuse and had not been proven toxic in humans." Nevertheless, the DEA overruled the judge and placed MDMA on Schedule I.

Lester Grinspoon, a professor of psychiatry and a psychedelic drug expert at Harvard, soon filed an appeal with a Boston court. The court ordered the DEA to "reconsider its decision because the issue of a potential medical use had not been appropriately considered." The DEA held firm. Even medical research on the drug would now require special FDA (Food and Drug Administration) approval.

Some researchers did try to get this approval. According to the German psychiatrist Torsten Passie, five applications to do research with MDMA were submitted to the FDA between 1986 and 1988.

"All five were routed for review but placed on clinical hold, effectively blocking them." The FDA based its reasoning for rejection "on the hypothetical risks of MDMA neurotoxicity." Hypothetical, because proper research had not been done. It was a catch-22.

As a result of this apparently ham-fisted ban, much of what we know about the effects of MDMA—changing only recently with the PTSD trials—comes from studies of recreational users. These are people who take the drug, often at the same time as other illegal substances, without the support of a trained professional, without always being able to ensure that the substance is pure, and without clear guidelines on how to use the drug in a way that minimizes risks. This introduces major confounds into the research. And it really is a sorry situation: it is well established that simply banning drugs does not make people stop using them; but it does drive their use underground. This in turn makes it that much harder for scientists to get a handle on the true benefit-to-risk profile that would apply to more controlled, therapeutic applications.

There are still therapists who conduct MDMA-enhanced sessions with romantic couples. But they, too, get the drug illegally and carry out their practice underground. This is not the way clinical research should be happening. It leaves us with little more than auspicious-sounding anecdotes from people willing to share their stories.

One such story comes from a longtime psychedelic therapist who shared her experience, anonymously, with UC Berkeley's *California Magazine*. We'll call her Carol. Carol says she has "safely and effectively" used MDMA with "dozens" of couples, spread out over the course of her thirty-year career. During this time, she and her husband have trained a whole network of therapists in their approach. In the interview, she explained how the drug can help her clients become "less defensive and less likely to project past traumas onto their partners."

"It's not always sweet and schmoozy," she said. "They often have difficult issues and come to me to look at the dynamics of their relationship. Who has the power? Who wants sex more than the other?

Should they separate? It's not always pretty, but it's alive. Seeing the other partner express fears can create more honesty in the relationship, and more intimacy."

Carol's protocol begins with separate MDMA sessions for each partner. These can last from three to four hours. After the groundwork has been laid one-on-one, there's a joint session for the couples to reflect together and decide what issues to address. Near the end of this joint session, as the effects of the MDMA begin to wear off, the partners work with the therapist to create a plan for how they are going to incorporate what they learned from the session into their lives. This is the integration we talked about earlier.

Some people might be concerned that swallowing a pill to achieve insights into one's relationship would be in some sense too easy. The sort of thing that, quickly attained, could just as quickly be lost. But thinking of the pill as an adjunct to relationship therapy should help alleviate this concern. It leaves plenty of room for active, nonsuperficial engagement and intentional learning about oneself and one's partner. As Carol describes the wrap-up to her sessions, the question is, "How are they going to follow up on these insights? They make decisions right there."

Drugs and authenticity

Unlike some other drugs, including psilocybin, which is also used by some underground therapists, MDMA doesn't cause hallucinations or seriously warp a person's sense of reality, at least at lower doses. You don't go into mystical realms; you aren't visited by fantastical creatures; you don't get the feeling that the boundary between yourself and others is entirely melting away. "With MDMA, the ego structure is intact," Carol explains. "The person can reflect inwardly without seeing the walls moving."

Psychiatrist George Greer has a similar take. In a recent interview he said that MDMA puts couples in an altered state while maintaining their normal communication faculties. The difference is that couples "don't get into their normal knee-jerk reactions," he said. "Fear does not push the mind into a defensive posture." During sessions, "they learn to communicate in more honest, direct ways. Then they learn how to employ that skill later without the drug."

But there are still concerns about authenticity. The President's Council on Bioethics under George W. Bush, in its landmark report *Beyond Therapy*, put the general worry like this: we might succeed in easing suffering with medication or, in this case, improving a relationship with a biochemical agent, but only at a cost. Specifically, we risk "falsifying our perception of the world and undermining our true identity."

How might this apply to MDMA? The drug doesn't typically cause visual hallucinations, so "false perception" in that very limited sense wouldn't be a concern. But MDMA does temporarily change how a person feels and reacts to otherwise fear-inducing stimuli, and it can lead to longer-term changes in how people understand their own emotions and relate to those around them.

So the "true identity" or authenticity worry might be something more like this: perhaps MDMA makes people more approachable, more trusting, or more empathic than they "really" are (at least while the drug is active). And among other possibilities, perhaps it makes them more open and honest with their partner, more confident in their priorities, and more at ease with themselves than they "really" are (sometimes even after the drug has worn off).

But if that's the worry, it isn't obvious why we should be so concerned. A drug that made you constantly chipper in the face of genuine suffering, or unable to perceive injustice, or unaware of emotional manipulation, or unresponsive to important social cues would indeed be a serious problem. A drug that "took you over" and haphazardly rearranged your inner life, or made you completely unpredictable, or rendered your thoughts and feelings unintelligible to yourself or others would also be a problem. But a drug that preserves your sense of self, and clears away incoherence, and makes you *more* tuned in to your emotions and those of your partner—as MDMA at least sometimes appears to do—is a different story altogether.

Of course, the way our emotions work *without* the use of a drug is sometimes inauthentic: the result of defense mechanisms, insecurity, anxiety, past traumas, or simply being extra hungry or tired or stressed out. The example of PTSD illustrates this principle very well. Consider the case of Jonathan Lubecky, an Iraq War veteran who used to fear Independence Day celebrations every year after his return home from fighting to South Carolina.

"Where I live, they love fireworks, so my whole neighborhood turns into Baghdad," he told a reporter. One night his wife walked into their bedroom closet to find him in full body armor, clinging to his service dog. He was having flashbacks. The strain these kinds of episodes would have on a relationship is not hard to imagine.

That experience of fear was real to Lubecky. It wasn't some bizarre self-deception. Moreover, it makes perfect sense why he would feel that way given all the horrors he had been through in the war. And yet in his current situation the fear was inappropriate: it was out of tune with the celebration that was happening in his actual environment. In other words, it was separating him from the reality of the world outside his head and putting up barriers between him and his wife. So what could be done?

In the eight years since coming home from Iraq, Lubecky tried to kill himself five times. He was taking more than forty pills a day to control his depression. All the while he was in psychotherapy. But it wasn't working. "When I was doing regular therapy," he says, "I wouldn't talk about the trauma because I felt like I was back there again. My body would react like it did when the trauma first occurred."

Lubecky was one of about two dozen war veterans, firefighters, and police officers who received MDMA for PTSD in the 2018 study we mentioned earlier—the one conducted by Michael and Ann Mithoefer and their collaborators. A month after the final MDMA session, he was one of the 68 percent of patients who no longer met the criteria for PTSD. As he told a reporter, he only found out about the trial when an intern working for his psychiatrist suggested that he google "MDMA PTSD." He had attempted suicide so many times at that point that he felt he had nothing to lose.

At the start of the first session, Lubecky had his own concerns about authenticity. He thought "it was going to be like in *The Matrix*, red pill, blue pill, kind of thing." But nothing at all happened for the first forty minutes after he took the MDMA. Then the drug kicked in and his therapist started asking questions.

"What's the weather like in Iraq?"

Normally, this is when Lubecky would shut down completely. But compared to his previous encounters with therapy, he says he was now "in such a safe place with the MDMA." As he described his experience, "I was able to just talk about things I had never talked about before without having a physical response to it." It was like "doing therapy and feeling like you're being hugged by everyone on the planet who loves you, and a load of puppies are licking your face."

Does that resolve the concern about authenticity? In one sense, perhaps it doesn't. After all, there *wasn't* a load of puppies licking Lubecky's face, and he *wasn't* being hugged by everyone on the planet who loves him. That's just how he felt while he was under the influence of a drug. But that's the thing with analogies: you aren't supposed to take them literally. If Lubecky had been licked by a load of actual puppies, he could just as well have said that it was "like being on MDMA." And he'd be right. The chemical reactions going on in his brain in both cases would be functionally similar, at least along certain dimensions, even though the trigger for the reaction in each case would be different.

Now if you found yourself habitually turning to drugs instead of real hugs and puppies to open up emotionally, this could be a problem. But the MDMA experience is pretty intense, and most people don't want to repeat it very often. Nor should they be able to, if future policymakers get it right. As Michael Mithoefer has stressed in interviews, MDMA isn't "something you should be able to pick up at the pharmacy and take home." Instead, he thinks—and we agree—that it should only be available through licensed clinics.

Part of the reason for this measure of control, apart from its use in guaranteeing that the drug would be pure and that any medical problems could be properly attended to, is that the likelihood of abusing MDMA following such clinical use appears to be low. In the trial Lubecky participated in, for example, six of the twenty-six participants had previously self-medicated with MDMA, illegally, from two to five times. In the twelve months after the trial, just two of those six participants took MDMA again, and none of the other twenty-four participants did. So under controlled conditions, when the drug is most likely to be effective, there appears to be less of a perceived need to keep using it on one's own.

As for Lubecky, the MDMA sessions have left him better off. His overall healthcare costs have dropped significantly, because his "pill count has gone from 42 down to 2." One medication he still takes is for a traumatic brain injury he sustained in Iraq. The other is an occasional dose of Ambien (zolpidem) to help him sleep. But he says that the changes in his personal life "have been even more profound." He is now a better father to his stepson and a better husband to his wife. Having harnessed the effects of a powerful drug, in concert with psychotherapy, his emotional life is now *more* authentic by any measure, and his relationships are on a much stronger footing.

Subclinical suffering

Does a person have to have PTSD for a drug to increase authenticity rather than diminish it? Some of the stories we shared earlier in this chapter suggest not. In general, there is a gradient transition between diagnosable mental illness and "ordinary" mental hang-ups that may distort our reality or cause us to frame things in unproductive ways. So there is reason to think that MDMA could help a larger group of people than those who are currently eligible for clinical trials.

Nevertheless, there are risks of driving a wedge between yourself and a stable, appropriately grounded version of reality. Many people have had the experience of feeling a certain way under the influence of a drug, only to later realize that they've made a horrible decision based on a mere perception that can't be sustained. This is the kind of thing that can happen outside of a therapeutic setting, where you don't have proper support to process your altered experience and emotions and put it all back together.

In the context of a relationship, for example, one possibility is that MDMA could conjure up feelings of love and attachment that are not properly rooted in an established connection. In fact, the journalist Mike Sheffield reports that in the 1980s recreational MDMA users referred to "instant marriage syndrome," leading them to create "shirts and bumper stickers advising people not to get married for at least 6 weeks after falling in love during an MDMA experience."

Reflecting on this advice could take us in a few directions. On the one hand, it seems to point to a recognition that MDMA-inspired love might not be the real deal after all, so you should wait for the drug's effects to wear off and really get to know a person before you make an important life decision like getting married. On the other hand, falling in love with someone *without* the use of a drug—or with a more familiar drug like alcohol—can similarly blind people to certain incompatibilities, leading them to rush into marriage while on a natural high only to confront a bunch of problems later on. It isn't clear then that the cases are so radically different. In fact, the takeaway for both scenarios might just be that making long-term, binding decisions while in the grip of extreme passion, drug facilitated or otherwise, is probably not a good idea.

Another direction you could go is to argue that love drugs like MDMA should only be used to help maintain or restore an existing bond—one that is already founded on an authentic connection between partners—while their use to spark new love with someone who isn't very well known should be strictly avoided. The thought here is that if you are already in a relationship with someone, and you have had time to consider your shared values, the strengths and weaknesses of your partnership, and the pros and cons of trying to improve your relationship with or without drug-assisted psychotherapy, then there would be less risk of making unrealistic or inauthentic decisions.

We agree with this conservative perspective and we suggest that initial research in this area should focus on existing couples. Yet in the long run, things may not be so simple. Even if a relationship starts with an inauthentic falling-in-love, an authentic love may still develop over time as shared interactions, conversations, and experiences combine to build a unique foundation. This is similar to how arranged marriages, marriages for wealth or convenience, marriages for citizenship, or any number of marriages not initially inspired by love can sometimes produce stable and caring partnerships over time that are just as authentic, by all appearances, as so-called love matches.

In arranged marriages, for example, the individuals do not first fall in love and then decide to spend their lives together. Instead, their families set them up on the basis of perceived compatibilities or other factors; it is then up to the couple to create love together as best they can. Of course, arranged marriages are widely variable. When a very young girl is paired with a much older man, or when couples are paired against their will, the marriage itself is morally objectionable and the background social structures are likely in need of reform. These could be called forced marriages, and they are wrong. But in some matches the pair is on roughly equal footing and has sufficient autonomy to go ahead with the marriage or not. And although they may not currently love each other, they may hope that, one day, they will.

Of course, you can't just will yourself to fall in love with someone. But as we've been discussing, you can certainly take steps, whether supported by drugs, environmental manipulation, or other techniques, that can make it so love is more likely. At the end of the day, we expect that the authenticity of a loving relationship will have more to do with the emotional dynamics that develop between partners, stemming from their shared experiences, than with the origin of the love whatever it may have been.

The question of identity

Another concern about authenticity is that drug-based interventions could introduce mental or behavioral inconsistencies in the person or couple being enhanced. A partner might start feeling or acting in ways they just wouldn't have before starting on the drug, and this could interfere with the sense that it's really the same person (or relationship) through time. Whether inconsistencies like this would really be introduced is not clear. It's an empirical question, but based on the evidence so far, it doesn't seem likely. MDMA, at least, doesn't pull on puppet strings and it doesn't commandeer your brain forever. Its immediate effects are temporary. The more lasting changes it can help to produce are available to your conscious reasoning and critical reflection.

We recently did some research on people's intuitions about potential changes in personal identity as a result of taking drugs. What we found is that if taking a drug leads to an improvement in your moral character (and perhaps also your relationship, though we haven't tested this yet), this change is likely to be seen as more identity-preserving than if your character deteriorates, even if the deterioration is due to going off a drug (like when a person stops taking their medication).

As an analogy, if you started feeling and acting more loving toward your partner after drug-free talk therapy (a relationship improvement), no one would accuse you of being inauthentic, much less a "different" person (except perhaps figuratively—and they would mean it as a compliment). But if the limited use of a drug in a therapeutic setting, or the lessons and habits learned from this use, also helped improve your relationship, it is hard to see why a different judgment should apply.

Close monitoring of any changes that do occur will be essential. The philosopher Marya Schechtman has suggested that authenticity and continuity in personal identity can be achieved if there is "empathic access" between the self that exists after a transformative experience and the one that existed before. This is a situation where your original preferences, beliefs, values, and desires are still present in the version of yourself that is psychologically altered. They still influence your personal experience and decision-making, even if they recede into the background to some extent during the process of alteration.

If you can recall and connect with how you felt (and what you believed) before taking the drug, and there is sufficient overlap and coherence between the new you and those past states and recollections, then according to this view you are still the same person as you were before. In other words, you are still your authentic self, despite having undergone some changes. At the level of a relationship this might mean that a couple's earlier love and regard for each other can be recalled with empathy and recognition during and after a neurological intervention. More generally, we shouldn't be so thrown off by the fact that a drug was involved in someone's process of change. We, and our relationships, are changed by life experiences all the time. An intense fight, a revelation of cheating, the birth of a child, a special vacation: all of these things can make us see ourselves and our partners in new ways, both positive and negative. Human psychology is remarkable in its elasticity—in the extent to which it can accommodate new experiences and integrate these into the self. Loving relationships can, and should, be elastic too.

Slowing down

We want to end on a note of caution. The stories we have been sharing so far are mostly anecdotal, and they have been biased almost entirely toward the positive end of the spectrum. Some people have had horrible experiences with MDMA, particularly in recreational contexts and with repeated heavy use. Some have died from overdosing, while others have had serious neurotoxic effects. However, let us not conflate illegal Ecstasy purchased from a dealer and taken while dancing all night in a hot warehouse, and pure MDMA taken once or a few times in a clinical, therapeutic setting with appropriate supervision.

"Recreational Ecstasy is often taken with heaps of other drugs," said Dr. Ben Sessa, a British psychiatrist and author of *The Psychedelic Renaissance*, a new book on the recent science of psychedelics. We caught up over Skype. "That includes alcohol and cannabis," he continued, "and opiates and benzodiazepines and cocaine and amphetamines. But mainly when we give people clinical MDMA, the patients are thoroughly screened. During the course of the MDMA experience, they'll have all of their vital signs monitored, like their temperature, blood pressure and heart rate, so they don't go wildly off. And crucially, after the experience they will have a lot of support for integration of the material that emerges."

When people take drugs recreationally, he told us, "they tend to not have any of those safeguards. They take an impure drug, they take it in non-facilitated circumstances, and they don't do any facilitation and integration afterwards of the work that comes out. So we mustn't make any inferences about the risk of MDMA based on recreational Ecstasy."

Sessa admitted that even in a clinical setting with pure MDMA there will be certain changes that might sound alarming, like elevated temperature, heart rate, and blood pressure. But these tend to be predictable, he added, peaking around forty-five minutes after taking the drug, staying at this peak for about four hours, and then coming down. And in a clinical setting all of these factors can be monitored and addressed; not so in "the sweaty nightclub where you don't chill out," as Sessa put it. The heat and lack of monitoring in a nonclinical setting increases the risk of negative effects on the brain as well as problems with water balance, which can lead to hypothermia or hyponatremia (depending on whether you drink too much water or too little).

In a recent article, Sessa emphasized that MDMA, when used irresponsibly, "has certainly been shown to cause physical, psychological and social harm, and even deaths. So we must be cautious and not disregard the concerns of those people who fear that medical use of MDMA may cause greater social and health problems than it may solve." Even so, he concludes that the emerging research pointing to a number of potentially positive uses for MDMA in contemporary psychiatry and psychotherapy—including the treatment of PTSD—suggests that the "evidence against *at least researching* MDMA" for therapeutic purposes "appears to be very scant indeed."

We agree and we want to amplify the call for research with this book. But we think the call needs to be widened from individual-level research focused on treating psychiatric disorders like PTSD to research that includes a relational aspect from the start. What kinds of couples could benefit from MDMA-enhanced therapy, and which couples should avoid it? Does there need to be a certain level of trust between partners to make such therapy worthwhile? Could this therapy be bad for some relationships, bringing up issues that were better left unexplored? The people asking these questions currently are underground therapists and recreational drug users collecting "data" informally. Better research is possible, and for the sake of our relationships, better research is necessary.

CHAPTER 7

EVOLVED FRAGILITY

POOR STELLA AND MARIO. As you may remember, things are not going well for them. She is miserable; he is miserable. There are many ways to describe their situation. The romance is fading. Real life is setting in. "That is what married life is like," one friend told them. "The problem is that you went into it thinking otherwise. Life isn't a Disney movie! You just need to change your expectations."

When we first introduced you to this couple, we described their deteriorating love in commonsense, everyday terms. We focused on their feelings as consciously experienced. But there is another way to describe what's happening to them, if we allow ourselves a peek under the proverbial hood. They are suffering a breakdown of their pair bond, part of the attachment system we have discussed.

Making sense of this breakdown is important for our project. It can help us see what's at the root of so much marital discord, going beyond the specific feelings of any one couple. We've said that practical ethics is about context and details, and it is. But the context of a problem can range from the local and particular to the vast and general. It's really the interplay between these levels of analysis that drives responsible and informed ethical thinking.

What is a pair bond, exactly? It is the powerful but limited grip of attachment, etched into our brains by the forces of evolution, that holds mammals (including humans) together while they rear their young. It wasn't "designed" for the modern world. It wasn't "meant" to last for a lifetime. Instead, the adult romantic pair bond is an intrinsically limited baby-raising mechanism we have inherited from our distant ancestors. Ancestors who lived in an alien environment compared to the one we now inhabit, and for whom mating and childrearing were very different endeavors.

This fundamental mismatch—between the underlying biological supports for love that are present in all of our brains and bodies, and the high-stress modern world we have created for ourselves through culture and technology—goes a long way toward explaining the fragility of contemporary relationships. The idea is fairly straightforward. Our nature as a species of animal, including our mating behavior and the feelings of love such behavior often trails or inspires, came about through a process of evolution by natural selection that took eons. *Homo sapiens*—that's us—has been in existence for a couple of hundred thousand years. Yet while our biology and deep dispositions have remained relatively stable over that time, the last ten thousand years have seen our world changed dramatically through agriculture and urbanization. And over the last two hundred years, technology has completely revolutionized how we live and love, and how we will be able to live and love in the future.

What came first, love or babies?

When a child asks, "Where did I come from?" there is a cliché response that starts, "When two people love each other very much . . . " It's a cute answer and it probably serves its purpose well enough, but in evolutionary terms it gets things almost exactly backwards. Babies didn't come about because our ancestors loved each other. Rather, our ancestors evolved the capacity to fall in love because they were already making babies—and love helped those babies to survive.

Until the last one hundred years or so (a blink of the eye from the perspective of evolution), infant mortality was very high. Indeed, in all primates, including humans, mortality is highest around birth and decreases as infants grow up and become stronger. If ancestral parents were to split up soon after a baby was born, this would put the baby's survival at risk. Human babies are particularly vulnerable because we are born at an earlier developmental stage than many other animals, so that care from both parents, at least in that ancestral environment, is thought to have been relatively more important.

In other words, it would have been in their own genetic interests for ancestral parents to stay together at least until the baby's healthy development was secure. This is because members of our species that formed a parental attachment (a pair bond) with each other during the most sensitive window for child survival would be more likely to transmit their genes to future generations—including the genes that favored such attachment.

How wide was that "sensitive" window, on average? Obviously, we can't go back in time to directly measure it in our ancestors, so we have to make inferences based on contemporary societies that are as close as we can get. To that end, modern hunter-gatherers such as the !Kung of southern Africa—whose survival and reproduction strategies are sometimes thought to be the nearest available approximations of those of our ancestors—have yielded some insights here. Among the !Kung, mothers typically breastfeed their babies for about four years. It has been speculated that this was the average gap between births for most of human history, assuming a general scarcity of resources.

In the late 1980s, the anthropologist Helen Fisher collected divorce statistics from fifty-eight societies. She found that when committed partners split up, they tend to do it around the fourth year of their relationship, typically after having had a single child. One interpretation of this finding—albeit a controversial one—is that couples tend to remain together for the minimum time necessary to raise one child together. After that, whether you stay together is of much less concern to evolution, so the pair bond may have a tendency to slip. Of course, there are countless exceptions to this tendency; that's how tendencies work. Many people choose and find ways to remain together until the end of their lives, and some of these people claim to be just as in love with each other as they were in the beginning (though this is not terribly common). Others remain together until their children have finished schooling.

There is a lot more to whether a romantic attachment lasts or fades than the evolutionary logic of our inherited capacity to form pair bonds. We will focus later on some of the social and historical processes that play into that trajectory. At the same time, we suggest that built-in biological limitations to lifelong bliss should not be dismissed as one important factor. As recently as 2016, Fisher has affirmed that "the three- to four-year divorce peak has not substantially altered over the past six decades, despite massive societal changes."

Another mismatch

Modern love, we claim, is saddled with old-world biological baggage. There is an analogy here with obesity. In his opus *The Omnivore's Dilemma*, Michael Pollan argues that at least one contributing factor to the modern obesity epidemic is the mismatch that exists between our evolved hunger drive, which is voracious, and our modern food environment, which is a veritable smorgasbord of unhealthy options:

> Human appetite [is] surprisingly elastic, which makes excellent evolutionary sense: It behooved our hunter-gatherer ancestors to feast whenever the opportunity presented itself, allowing them to build up reserves of fat against future famine. Obesity researchers call this trait the "thrifty gene." And while the gene represents a useful adaptation in an environment of food scarcity and unpredictability, it's a disaster in an environment of fast food abundance, when the opportunity to feast presents itself 24/7. Our bodies are storing reserves of fat against a famine that never comes.

Our ancient pair-bonding instincts may be similarly mismatched to this Tinderized, digital age of romantic distractions, with its abundance of possible partners and lifestyles. As we have more and more options for potential mates, our psychological limitations lead us to believe that there will always be someone even better "out there."

Consider *Playboy*, whose editors reportedly sift through about six thousand pictures of models for the monthly centerfold. From these a tiny handful of photos is selected, and even those are carefully retouched before publication. What *Playboy*-purchasing men in contemporary societies see, then, "are the most attractive women in their most attractive poses with the most attractive background in the most attractive photo-shopped images." This is in blatant contrast to what our distant ancestors would have experienced, living in groups of around 150 people, with no photography or magazines.

This contrast between their world and ours may have real-life consequences. After viewing images like these, men are less happy in their relationships and more prone to looking for other options. Heterosexual women are adversely affected, then, not only because their partners are less likely to be satisfied with their appearance, but because the pictures inspire an unhealthy competition among women to look more like models.

Throw in other modern novelties like ease of long-distance travel, the ballooning size of social groups, the decoupling of sex from reproduction due to birth control, and the ready availability of management strategies for sexually transmitted infections (safer sex, cheap testing, antibiotics, and antiretroviral drugs), and you get an increase in opportunities for relatively low-risk love affairs. For monogamous relationships, at least, this can put still further strain on the tenuous bonds of attachment.

In short, in our world—a world we have created for ourselves with technology, culture, politics, advanced social institutions, globalization, and other modern forces that move a lot faster than evolution by natural selection—adult romantic pair bonds have a troubling tendency to wear themselves out, or shatter, long before "death do us part." Divorce has now overtaken death as the major cause of marital breakup.

Even when pair bonds don't break, they can pose other challenges. For example, they can frustrate our need to feel free and independent. As somebody once said, "There are two great tragedies in life, losing the one you love and winning the one you love." Our need for security propels us toward committed monogamy, and our need for adventure tends to pull us in the other direction. Yet modern romance promises that we can have it all in a single perfect relationship that satisfies our romantic, emotional, and sexual needs. Once we realize how implausible that promise is, it's not surprising that so many relationships crumble in disappointment.

Evolution did not "design" us to have lifelong monogamous marriages in the modern world. It didn't even design us to be happy. What it did was give us a strong desire to mate with each other, not necessarily exclusively, and to stick around in the case of children long enough for them to survive and find mates of their own. As the anthropologist Donald Symons has succinctly put it, natural selection promotes "reproductive, not marital, success."

Marriage and divorce

Divorce rates are notoriously hard to pin down. If you've ever wondered where that famous "half of all marriages will end in divorce" statistic comes from, here is what we could find. It comes from the number of divorces per 1,000 people in a given population—typically the U.S. citizenship, based on census figures—divided by the number of weddings per 1,000 people in the same population over the course of a year. This ratio, at least in recent decades, has remained at or near 1:2 across a range of samples, hence 50 percent. Basically, there are about twice as many weddings as there are divorces in a given population in a given year.

Why so many divorces? Limitations in our mating, bonding, and childrearing biology—the evolutionary mismatch we've been describing—are likely to be at least part of the problem. But as we said, they are not the whole story; when it comes to assigning blame for the fragility of modern relationships, social, historical, and even economic factors deserve their fair share.

The best way to see this is to consider how relationship norms have changed over the past 150 years or so, including a whiplash-inducing transformation of the institution of marriage. Until very recently, marriage was not primarily based on love. Passion, at least, was often found outside of wedded bonds, and marriage was more about connecting families and maintaining wealth or the capacity for labor. In other words, right up until the Industrial Revolution, marriage accomplished much of what we expect markets and governments to do in modern societies.

To be sure, people fell in love—much as people fall in love today—before the Industrial Revolution. But marriage was not *about* being in love. It was too important an economic and political institution to be entered into on the basis of unstable romantic feelings. As the foremost historian of marriage Stephanie Coontz has put it, "For most of history it was inconceivable that people would choose their mates on the basis of something as fragile and irrational as love and then focus all their sexual, intimate, and altruistic desire on the resulting marriage."

But that was then. Today, far from being inconceivable, marrying for love is, in the West at least, pretty much the only game in town. Four factors have led to this transformation:

1. Men and women are no longer understood to be fundamentally different sorts of creatures (it was seen as natural and inevitable, for example, that women—in contrast to men—had no sexual desires). Now mutual sexual fulfillment is expected in relationships.
2. Relatives and community members have less influence in shaping individuals' life courses, owing to urbanization, increased social and geographic mobility, and the development of economic interests that put more weight on a person's resources and credentials than on who they're married to.
3. As we mentioned, the invention of reliable birth control and methods of reducing the risk of sexually transmitted infections: women are now relatively less fearful of unwanted pregnancy, which has increased

their sexual freedom. And abolishing the legal category of "illegitimacy" has weakened the relationship between socially acceptable childbearing and marital status.

4. Women are no longer (as) legally and economically dependent on men, nor are men "domestically" dependent on women (although women are still more likely to be burdened with the majority of housework and childcare).

As these various paths to increased gender equality and personal autonomy opened up, it became harder to pressure people into marriage—or into staying married come what may. People no longer had to get married in order to build a functioning, meaningful life for themselves. If they wanted a long-lasting sexual and emotional partnership, marriage or no marriage, they could form it with whoever they personally desired—liked, got along with, *loved*—clearing center stage for romantic passion to play its leading role in modern committed relationships.

The paradox of love

But is love up to the challenge? We seem to have created a paradox. On the one hand, we are now free to pursue our personal projects, seek sexual gratification, and escape the burdensome strictures of the past. Husbands can no longer legally beat their wives; children have a right to go to school (instead of being forced to provide free labor for their families); sexual double standards have been reduced, if not eliminated. On the other hand, some argue that trailing behind these moral advances are important losses. Traditional structures and forces like religion, extended families, and communities did not just limit our freedom, sexual and otherwise. They also provided a sense of belonging, social support, order, and meaning. By dismantling them, we have given ourselves more options and fewer constraints, but have also made ourselves more alone.

That is the central paradox of modern love—and it's love that we expect to solve it. Going back in time is not an option. As Coontz notes, we can no more "turn back the clock in our personal lives" than we can "go back to small-scale farming and artisan production" in our economic lives. The central role of love in contemporary relationships is here to stay. It's tied up with so many of our other values. In their book *The Good Marriage*, authors Judith Wallerstein and Sandra Blakeslee capture this notion perfectly:

> In today's world it's easy to become overwhelmed by problems that seem to have no solution. But we can shape our lives at home. . . . The home is the one place where we have the potential to create a world that is to our own liking; it is the last place where we should feel despair. As never before in history, men and women today are free to design the kind of [relationship] they want, with their own rules and expectations. . . . In our fast-paced world men and women need each other more, not less. We want and need erotic love, sympathetic love, passionate love, tender, nurturing love all of our adult lives. We desire friendship, compassion, encouragement, a sense of being understood and appreciated. . . . We want a partner who sees us as unique and irreplaceable. . . . A good [relationship] can offset the loneliness of life in crowded cities and provide a refuge from the hammering pressures of the competitive workplace. It can counter the anomie of an increasingly impersonal world, where so many people interact with machines rather than fellow workers.

In short, love, and loving relationships, have become the glue holding many of us together as we face the storms of modern life. If we want this glue to last, we need to understand its complex nature, starting with the basic pair bond that forms between romantic partners. One neurochemical in particular has been hypothesized to strengthen this bond directly, so we decided to give it its own chapter. We are talking about oxytocin.

CHAPTER 8

WONDER HORMONE

FOR ALMOST A DECADE, the hormone oxytocin has been relentlessly hyped as a "one-ingredient recipe for a utopian society." In the words of one of our favorite science writers, Ed Yong, this "molecular high-five, which is released when we hug, tweet, dance, and orgasm, has been linked to trust, cooperation, empathy and a laundry list of other virtues." One online magazine has called oxytocin "the most amazing molecule in the world." Other popular accounts have taken an alliterative turn, calling oxytocin a "cuddle chemical," "hug hormone," or "moral molecule." And yes, some have even called it a "love drug" (no alliteration there, but closer to our hearts).

On Amazon you can buy a nasal spray called OxyLuv. The spray, which purports to improve your sex life while decreasing your anxiety, has earned three out of five stars for satisfaction—across an array of customer experiences. On the one hand, a Brian Williams (not, we presume, the erstwhile host of *NBC Nightly News*) claimed that the spray made him more "willing to engage," but only when he poured it into his morning coffee. On the other hand, a self-declared researcher from the University of Washington wrote that he had analyzed the spray with two separate tests and found it to be "negative for oxytocin" and "consistent with tap water."

Still, a range of products touts some version of oxytocin, including Oxiboost, Liquid Trust, and Attrakt Oxytocin Pheromone Spray Cologne for Him, which promises to "create lasting attraction" and "enhance existing relationships." In other words, Amazon is peddling real-life love drugs, although, most likely, not very effective ones.

Nevertheless, these products are attempting to capitalize on research into actual oxytocin, a hormone that *is* involved in human attachment, as we have seen. But oxytocin doesn't have all those mystical properties that have been attributed to it. In fact, it provides a very good illustration of just how complicated and messy designing any new real-life love drug will be.

Science time

Let's start with a little science. Oxytocin is what's called a neuropeptide, a protein-like molecule that neurons use to communicate with each other. It's present in a lot of animal species, from reptiles—in the form of something called vasotocin, an evolutionary precursor—to mammals, including us primates. And it is intimately associated with social, sexual, and reproductive behaviors, such as mating and childbirth.

The part of the brain that produces oxytocin (OT, as we'll say from now on) is called the hypothalamus, and it releases OT when it's called for various jobs. For example, OT sometimes acts as a hormone, controlling functions like labor contractions, the release of breast milk, and cervical dilation during childbirth. But OT isn't just for mothers and birthing. It also acts as a neurotransmitter in everyone when it's released into the central nervous system, where it influences a whole slew of interpersonal judgments, motivations, and behaviors. It may not be a "moral molecule," but it is a multitalented molecule to be sure.

Much of what we know about OT comes from studies of voles. Voles are small rodents that, for our purposes anyway, come in two main varieties. Depending on the species, they follow either a monogamous or a polygamous mating strategy, and the difference appears to turn on the activity of OT, vasopressin, and dopamine and how their respective receptors are patterned in the brain.

The basic idea is that oxytocin and vasopressin help register and store information about social identity (so you know who your partner is) and then dopamine associates that information with some level of reward (so you feel motivated to spend time with your partner). Roughly speaking, the denser the cluster of OT and vasopressin receptors in the brain's reward centers, the higher the social reward and the more exclusive the partner preference.

Prairie voles have a "high density" cluster and they are usually socially monogamous. Meadow voles have a "low density" cluster and they are usually polygamous. Critical studies have manipulated OT and vasopressin levels directly in the monogamous (prairie) species. Infusing oxytocin into the brains of the females and vasopressin into the brains of the males facilitated a pair bond even in the absence of mating. Normally the voles have to have sex for this to happen. In other studies, infusing an oxytocin blocker into their brains had the opposite effect: the voles didn't form a partner preference even after mating (whereas normally they would).

To be clear, no one knows if human attachment relies on the exact same hormonal machinery as that seen in voles. But several researchers have argued that natural selection would be under pressure to conserve a biochemical system that is so central to the survival and reproduction of other mammals.

Neuroimaging studies in humans seem to support this. Mothers who observe a photograph of their own young child, compared to another young child they have known for about the same amount of time, but did not give birth to, show heightened activation in brain areas rich in OT, vasopressin, and dopamine receptors. This effect is amplified for mothers with a secure attachment style compared to those who are insecurely attached. And it correlates with OT levels in the blood. Similar activation patterns happen in adults exposed to an image of their romantic partner—with whom they report being "intensely in love"—compared to an image of a platonic friend.

Although none of these findings demonstrates that elevated OT levels in humans directly *induce* greater feelings of love or attachment—whether in the mother-child or romantic partner case—they do point to a clear *correlation* between OT and attachment that seems to mirror what has been observed in other species.

Correlation does not entail causation, as everyone knows. So recent experiments have gone even further and have attempted to investigate a potential causal relationship between OT and various phenomena associated with love and attachment.

Here is how these experiments basically work. Using a simple nasal spray to deliver OT into the brain through one nostril, these studies were placebo-controlled and double-blind. This means that neither the scientists nor the study participants knew who got the real OT versus a dummy spray until after the results were in. (This kind of design is usually considered the gold standard for assessing causation. On the other hand, it's also been called a "golden calf" because scientists are sometimes so in awe of it they forget to properly examine the actual study materials, statistics, or results.)

These foundational studies have delivered a range of important, if tentative discoveries. Emphasis on tentative. In just the last year or so while writing this book, we have seen evidence that some of the early findings on OT might not replicate. As one of the main researchers behind the famous vole experiments wrote in a recent review, "Intranasal OT studies are generally underpowered," meaning they have too few participants for the statistics to make sense; and, "There is a high probability that most of the published intranasal OT findings do not represent true effects." Accordingly, "the remarkable reports that intranasal OT influences a large number of human social behaviors should be viewed with healthy skepticism."

This a stunning admission for a subfield of science that has long been at the vanguard of the media-hype parade. Science, in contrast to the fast clip of that parade, moves slowly, and several major research fields have been grappling with how to improve their methods. This has become a big issue in light of the "replication crisis" that is now rippling through science and medicine. Brian has written a lot about this perceived crisis in a separate line of work. But in a nutshell, the normal ways of doing science over the past many decades—from initial data collection all the way through to peer review—have likely been churning out a lot of false alarms. Not all false alarms, but more than you'd hope. And it isn't just OT research; in fact, it's research in many different subfields of psychology, economics, biology, and other disciplines as well. What this means is that you should take what we are about to say with respect to specific OT findings with a decent-sized grain of salt. We think we have the bigger trends about right, but only time will tell which of the particular results we mention will stand up to independent replication.

The gist of what we know

From our reading of the literature, the balance of evidence suggests that externally boosting OT levels can result in prosocial, bond-enhancing outcomes for some individuals or couples; but it may also lead to antisocial, adverse outcomes for other individuals or couples—or the same individuals or couples under other conditions. In short, if OT, or any other potential love drug, is going to be useful as a relationship enhancer at some point in the future, it will likely only be so for some people under some conditions.

The most famous study in this area is probably one by the Swiss neuroscientist Beate Ditzen. In 2008 Ditzen and her colleagues administered OT (versus a placebo) to nearly fifty heterosexual couples before having them engage in a conversation about a chronic source of conflict—basically, having them start an argument. Interactions were videotaped and then coded for verbal and nonverbal communication behaviors, and the researchers took salivary cortisol measures as well (cortisol is a stress hormone). The headline finding was that OT increased the ratio of positive to negative communication behaviors and facilitated a more rapid reduction in cortisol levels after the conflict.

As Ditzen noted about her research later on, "This was the first study of its kind and important because it evaluated real-time natural couple behavior in the laboratory." OT, she said, "might help us to [enhance] the effects of a standard treatment, such as cognitive behavioral therapy, by possibly making the benefits of social interaction more accessible to the individual." She cautioned, however, that OT should not replace those well-worn treatments—which is the same caution we have been giving throughout this book.

More generally, OT administration appears to hold promise for boosting emotions or behaviors that are central to the health and functioning of relationships. Specifically, it has been shown—in a grain-of-salt kind of way—to reduce anxiety and stress; increase trust, eye contact, mind reading, and empathy; and make it easier to remember the good parts of relationships. In a recent study, OT enhanced the effects of partner support in reducing the subjective experience of pain.

In a relationship therapy context, these effects could make it easier for partners to approach each other with less defensiveness and take on the other's perspective (like the preliminary findings on MDMA we discussed before). If so, this would likely aid the therapeutic process. In another study, OT-primed male participants who were in a committed heterosexual relationship—but not OT-primed single males—kept themselves at a significantly greater distance from an attractive female experimenter during an initial face-to-face encounter. They also had less of a reflexive "approach" response when exposed to erotic images of attractive women on a computer.

If these findings hold up, they might suggest that intranasal OT could one day promote sexual fidelity toward a current partner. In line with this view, a study by Adam Guastella, a leading OT expert, showed that male participants were quicker to identify positive stimuli associated with bonding and relationships when under the influence of OT. More generally, OT may make the encoding of positive social cues more rewarding and memorable. As the neuroscientist Nadine Striepens and her colleagues have suggested, "By acting as a neuroplasticity agent, oxytocin may help rewire neural systems, so that specific cues from individuals with whom bonds are formed are more likely to elicit recognition and pleasure in the future."

The dark side

OT seems to have a dark side as well. Not all of its effects are pro-relationship. Researchers have provided evidence that in addition to promoting empathy and generosity, OT can increase feelings of envy and schadenfreude (the enjoyment of others' misfortune). It also seems that people behave more cooperatively under the influence of OT when interacting with familiar individuals, but less cooperatively when the person is a stranger. And there is even evidence that OT can increase in-group favoritism or ethnocentrism while at the same time potentially magnifying out-group prejudice. These results suggest that OT could ultimately turn out to be a double-edged sword: capable of increasing bonds among friends but also boosting fear and distrust of strangers.

Distinct mental traits have also been associated with negative effects of OT on behavior. People with borderline personality disorder, for example, may exhibit *decreased* trust and cooperativeness under the influence of OT. And people with an anxious or insecure attachment style appear to remember their own mothers as being less close and caring. Finally, OT seems to improve empathic accuracy only in less socially competent individuals, whereas it has no effect for those who are already well tuned to social information.

These and other findings suggest that the effects of OT are not straightforward. How it influences thoughts and behavior depends on the biological profile of the particular person, their psychology, and the social context in which they are acting.

Assuming that enough of these findings turn out to replicate, the next step would be to create an inventory of benefits and risks that would weigh in favor of, or against, the use of OT for particular individuals or couples. These factors would have to include psychological traits like bipolar disorder, borderline personality disorder, and insecure attachment styles; genetic profiles, including genes that govern the distribution of OT and vasopressin receptors within the brain; and relevant situational variables. Couples would have to be screened for any complicating factors before OT administration, and individuals or couples who were at a high risk for bad outcomes would likely need to be excluded from OT-enhanced therapy.

Some caveats

All of that being said, the potential for bad outcomes associated with OT administration has to be kept in context. Consider the finding that OT may increase feelings of envy (or decrease cooperation) when interacting with strangers. Since romantic partners are not strangers to each other, it seems reasonable to hypothesize that at least these types of negative out-group or stranger-driven effects would not be too likely in a counseling setting.

Besides, even if OT does promote such troubling outcomes as in-group favoritism (at the expense of more altruistic out-group judgments), this wouldn't be enough to rule out OT-enhanced therapy. First, you'd have to show that any undue negativity toward out-groups extended beyond the time frame of the counseling session in which the in-group therapeutic effects were being exploited. And second, you'd have to show that the bad effects were stronger when OT is administered through a nasal spray than when it's released into a person's brain by the hypothalamus. Obviously, couples who wanted to improve their relationship through intimate activities—say, sex—which normally trigger the brain's own release of OT, should not be expected to refrain from these activities simply because they might inspire a transitory in-group bias.

All of this shows how important it will be to study the effects of OT on human judgment and behavior outside of the lab. The rapid effects of a single dose of OT on sensitive outcome measures, which are typical in current research, need to be systematically compared against the effects of repeated use in more realistic settings.

The state of the art

Research in this vein is currently underway. At the time of writing, Adam Guastella and his colleagues are running a study to test the longer-term effects of OT on couples in therapy. As part of the study, couples are asked to discuss an issue that normally causes conflict and then try to resolve the issue. Guastella tells us that data analysis is still in progress, but he "expects couples that got oxytocin to show less hostile interpretations of the problem and be less critical of their partners." Based on previous findings, the hope is that there will be increased perspective taking and less blame, leading to more productive communication and joint problem-solving going forward.

Psychology professor Ruth Feldman, at Bar-Ilan University in Israel, who has spent years studying the role of OT on the mother-child bond, has commented on Guastella's study. As she sees it, it is plausible that OT and these relationship-supportive actions could form a positive feedback loop. "Oxytocin can elicit loving behaviors," she says, "but giving and receiving these behaviors also promotes the release of oxytocin and leads to more of these behaviors." She also thinks that while talk therapy on its own can bolster the OT system, this process could in some cases be helpfully jump-started by administering OT through a nasal spray. Assuming that Guastella's findings support his hypothesis, combining talk therapy with OT could be an effective way to improve communication between partners, particularly when a hostile or defensive communication style was learned in childhood and is still causing problems.

Ethical considerations

As the sort of therapy Guastella is pioneering progresses, it will be important for couples who want to improve their relationship with OT-infused treatments to do so under the guidance of a trained professional. This professional should be responsible for undertaking any prescreening requirements, as we noted above, and should make sure that careful monitoring of the relationship regularly takes place. Finally, both partners would have to give their informed consent to any drug-based relationship intervention, and either partner should be free to stop it at any time.

More specific cautions are also worth raising. OT can apparently increase a person's willingness to take risks in interpersonal relationships. It can also increase cooperation in some individuals by reducing their aversion to being betrayed. Both of these effects can be good, bad, or neutral, depending on the context. For example, within certain relationships, an instinctive unwillingness to suffer repeated betrayals of trust could be exactly what is needed to keep your partner from taking advantage of you. If OT undermined that unwillingness, it could spell serious trouble.

Users would need to be aware of these potential influences on their relationship. And they would need to make sense of them in light of their own values. The imminent prospect of neurochemical modification of human relationships calls for the development of more general ethical guidelines concerning when and how to roll out these new technologies.

Still, within the context of these more general principles, any OT-based treatment plan would have to take into consideration the specific contextual factors that apply to the unique situation of each couple. There are a lot of moving parts. For example, as we pointed out, while some couples might benefit from trust-enhancing interventions, other couples might be put at risk. In addition, although negative stranger or out-group effects would probably be less of an issue for romantic couples, it's at least conceivable that some relationships will have deteriorated to the point that the individuals do feel like strangers to each other. Future research should look at whether OT (or other drugs that affect the OT system, like MDMA) would have positive or negative effects in cases like that.

Finally, coercion is a red line. A particularly bad situation would be if one of the partners tried to cajole the other into remaining in the relationship against that person's wishes, whether by pursuing

OT-enhanced therapy or otherwise. Partners might also disagree on the extent to which they believe the relationship is worth strengthening. These kinds of worries can be allayed, in principle, by the conditions we spelled out earlier: the administration of OT, or any other potential love drug, should only be pursued if both partners autonomously agree to its use in the relationship; and such drugs should only be administered on the basis of careful ethical reflection under the guidance of a licensed professional trained in both couples counseling and the use of the drug in question.

The stress here is on "in principle." Some bioethicists argue that the cool-headed rationality required by conditions like the ones we have just laid out is not all that common in real-life medical decision-making, and may even be a myth. In real life, people make their decisions about therapy or other healthcare in a fog of desperation, confusion, and stress, while balancing all sorts of competing interests, from their own pain, discomfort, and fears to those of others. Romantic relationships may involve all of these pressures and more. Adding drugs to the mix will only make things more complicated. It will be crucial to get a handle on actual power dynamics and shifting contextual factors when bringing drugs into romantic relationships. This will be especially important for drugs intended not to restore love but to bring it to an end. We have been calling these anti-love drugs, and we turn to them next.

CHAPTER 9

ANTI-LOVE DRUGS

THE IDEA OF AN ANTI-LOVE REMEDY, a cure for love, is almost as old as love itself. References crop up in the writings of Lucretius, Ovid, Shakespeare, and many others and are tightly linked to the idea that love, desire, and especially obsessive infatuation can sometimes be like a serious illness: bad for your physical and mental health and in some cases destructive to your overall well-being. The playwright George Bernard Shaw called romantic love one of "the most violent, most insane, most delusive, and most transient of passions" and even mocked the idea that modern marriages should be based on so volatile and fleeting an emotional foundation.

Ancient cures for love, desire, and infatuation included phlebotomy or bloodletting, exercise, avoidance of rich foods and too much wine, and keeping properly hydrated. Or if *Harry Potter* is more your style, a love potion antidote can reportedly be brewed from Wiggentree twigs, castor oil, and the extract of a Gurdyroot. Although these examples are prescientific or fictional, they point to a conception of love as something rooted in the body—this is the biological dimension we talked about earlier—and so susceptible to being treated by the ministrations of a doctor (or wizard as the case may be).

Modern neuroscience goes a step further and traces the biological side of love to the brain specifically. In 2008 one of us (Julian) along with our good friend and colleague Anders Sandberg, published the first-ever discussion of the science and ethics of chemically enhanced love and relationships. As with the tale of Stella and Mario, this account focused on the potential use of biochemistry to help maintain a seemingly worthwhile relationship that might otherwise needlessly break down. The following year, writing in *Nature*, the neurobiologist Larry Young raised the possibility of pushing love in the opposite direction with a chemical cure.

As Young sees it, love is essentially "a cocktail of ancient neuropeptides and neurotransmitters." He argues that "drugs that manipulate brain systems at whim to enhance or *diminish* our love for one another may not be far away." Although we agree with Young's general point, his conception of love could use some refining. Given that love has a dual nature, as we outlined earlier, we think it would be better to say that the *biological* dimension of love is an emergent property of an ancient chemical cocktail, whereas the *psychosocial* dimension of love emerges from practices, cultural norms, and institutions embedded in societies. With this clarification in mind, what kinds of drugs could be added to the biological cocktail to bring about the effects Young speaks of?

Low-level candidates

Some candidates for anti-love drugs are already being tested, albeit informally. In some communities, Orthodox Jewish yeshiva students are being prescribed psychiatric drugs to suppress sexual feelings. Rabbis and marriage counselors are encouraging this so that the students can find it easier to comply with rigid Orthodox norms about love and sexuality. (We will talk about the ethics of this, or lack thereof, in a later chapter.) In another example, a Christian man suffering from what he described as internet sex addiction, a condition he felt was ruining his marriage, was prescribed oral naltrexone—a drug normally used to treat alcohol and opioid addictions—to control his urges. And American sex offenders are sometimes offered chemical castration through the ingestion of antiandrogen drugs as a condition for parole.

These cases involve the clumsy use of rather low-level pharmacology to target the bodily sex drive. In that sense, they are only capable of indirectly shifting a person's loftier feelings of love or attachment. Even so, they raise difficult, even disturbing moral questions.

We aren't expecting that some single pill will ever be invented that could make a person fall out of love with someone or break their attachment to an ill-suited partner. Different drugs might be needed for different situations. For example, a drug to dampen the urges of someone with pedophilia would likely be different from (and work differently on the brain than) a drug designed to help an abuse victim sever unwanted emotional ties to the abuser. Likewise, a love vaccine that works to prevent unwanted love from ever developing might differ in meaningful ways from a love antidote designed to quell an existing love, which might in turn differ from a memory-altering drug that could help someone recover from a prior love or past romantic trauma.

This last example is not actually speculative. Nor is it fictional, despite its resonance with the plotline of the romantic sci-fi dramedy, *Eternal Sunshine of the Spotless Mind*. In this 2004 film, a heartbroken man played by Jim Carrey undergoes an experimental treatment to erase memories of his ex-girlfriend, played by Kate Winslet, after learning that she has done this following their painful breakup. The concept behind the film apparently came out of a chance conversation between two of its co-writers, Pierre Bismuth and Michel Gondry. As Gondry recounts, one of Bismuth's best friends "was always complaining about her boyfriend, and he was tired of her whining about it, so one day, he asked her: 'Listen, if you had the opportunity to erase him from your memory, would you do it?' And she said: 'Yes.'"

The film was a huge success, both critically and at the box office. It seems the desire to erase a romantic partner from one's mind is one that many people can relate to. But using drugs or other technologies to meddle with memory, especially as a way of healing from a traumatic relationship, was not exactly feasible when the film came out. That is probably part of why the film was so successful: it depicts a world that is relatable and recognizably ours, but just beyond the reach of current science. As our book was heading to press, however, reports of a real-life technique for memory modification in response to heartbreak hit the stands.

The coverage centered on Alain Brunet, a psychiatrist and expert in PTSD at McGill University in Montreal, Canada. Healing from heartbreak, he claims, can sometimes be as hard as healing from violence-induced PTSD: "Greek tragedies have been written about it. It's not a banal incident. People often cite a breakup or a divorce as their worst life experience."

In his lab, Brunet works with victims of what he calls "romantic betrayals." These can range from harassment by a former lover to sudden abandonment by a long-term partner. He uses a technique known as reconsolidation therapy, which combines a drug-based treatment with practical exercises to change the emotional content of disturbing recollections. "We don't treat the symptoms," Brunet says of his method, "we treat the memory."

Unlike the fictional staff of Lacuna, Inc., the memory-erasure firm in *Eternal Sunshine of the Spotless Mind*, Brunet and his collaborators stop short of trying to "delete" traumatic memories altogether. "You don't forget your memory," Brunet stresses, asking, "Who would want to forget their love story?" (Presumably some people would; whether they should is another question.) Rather, the goal of the therapy is to keep the memory intact while removing its traumatic aspects. Here is how this actually works:

> An hour before a therapy session, the patient is given a dose—between 50 mg and 80 mg—of a beta-blocker called propranolol, and is asked to write a summary of the traumatic experience, following a strict format: a first-person text in the present tense that describes at least five physical sensations felt at the time of the event. By reading the summary out loud, the patient "reactivates" the memory, and does so over four to six weekly sessions, under the influence of propranolol. At every reading, the memory is "recorded again" while the drug suppresses the pain it contains.

The therapy has a good success rate. More than 70 percent of participants in a 2018 study conducted by Brunet and his colleagues reported meaningful relief from their breakup-related stress. Following treatment, many patients said reading through the details of their memory felt like "reading a novel." In other words, the narrative remained, but the pain was gone.

According to Brunet, both the drug (propranolol) and the writing exercise work together to bring about the effect: "The pill or the session alone will not work."

Three building blocks

What other cures for love might science soon deliver? To make sense of these different options, we will organize them around the three distinct brain systems that researchers argue form the biological building blocks of romantic love. As a reminder, these are *lust*, *attraction*, and *attachment*. Each of these is thought to serve a different evolutionary purpose, and each can (and does) function somewhat independently in humans and other mammals. In other words, it is possible to be attached to one person, attracted to someone else, and lusting after another. As Helen Fisher and her colleagues have put it, "Men and woman can copulate with individuals with whom they are not 'in love'; they can be 'in love' with someone with whom they have had no sexual contact; and they can feel deeply attached to a mate for whom they feel no sexual desire or romantic passion."

At the same time, the hormonal and neural circuitries involved in these three drives interact and overlap. For example, testosterone can stimulate the production of vasopressin; oxytocin can modify activity in dopamine pathways; and serotonin can alter the function of several other neurotransmitters. Accordingly, the basic building blocks don't always stack up into a nice, neat, well-designed, durable tower of love. They sometimes stack up into violent obsession, addictive attachment, pedophilia, erotomania, or allegiance to a dangerous cult leader. Although the developmental pathways, and obviously the end result, may differ, the same basic neurochemistry is involved in all of these phenomena.

Zach Lynch, founder of the somewhat ominously named Neurotechnology Industry Organization, predicts that sophisticated nanobiochips and advances in brain imaging will allow for the development of what he calls "neuroceuticals," or highly efficient synthetic drugs that could target specific areas of well-defined brain circuits. But this sort of finely tuned technology is only just emerging on the distant horizon; we have no idea how long it might be before Lynch's prognostications can be verified. Let's concentrate then on what is available here and now. Some of the drugs we will discuss in what follows make an appearance in more than one category—lust versus attraction versus attachment. And where we are being more hypothetical than descriptive, we will try to make that clear.

Anti-lust drugs

Let's start with anti-lust interventions. These are the easiest to pull off. Drugs acting on the lust system are definitely already available. We gave three examples earlier: the psychiatric drugs given to the yeshiva students were antidepressant medications; we also brought up androgen blockers for sex offenders, and oral naltrexone for the man who was sex obsessed on the internet.

We can also add a few household examples, namely tobacco and alcohol. And then there is a range of other drugs with libido-weakening effects among their potential outcomes. These include almost all blood pressure pills, pain relievers containing butalbital as well as opiates such as morphine and hydrocodone, statin cholesterol drugs, certain acid blockers used to treat heartburn, the hair loss drug finasteride, and seizure medications including gabapentin and phenytoin. With the exception of androgen-reducing drugs used specifically for chemical castration, the negative effects of these substances on a person's sex drive are typically neither intended nor desired. Yet as we saw with the off-label administration of antidepressants to the yeshiva students, that doesn't necessarily have to be the case.

What is the mechanism of these drugs? Basically, it is regulation of testosterone. Focusing on this hormone as one of the most important biological factors behind sexual desire and behavior, a number of studies have measured the effects of reducing testosterone on problematic sexual thoughts or activities—mostly in men—such as intrusive erotic fantasies or compulsive exhibitionism (like flashing random strangers in the park). One study, for example, reported that cutting testosterone levels led to a reduction in pedophilic sexual fantasies and urges among some men. Likewise, the neuroscientist Till Amelung looked at the combined effects of androgen deprivation therapy and group psychotherapy on a small sample of "self-identifying, help-seeking pedophiles," and reported a reduction of inappropriate sexual behaviors, an increase of risk awareness and self-control, and a dampening of thoughts that tend to correlate with pedophilic actions.

Side effects are a huge problem with these drugs. In one study, antiandrogen drugs were given to hospitalized patients struggling with a range of paraphilic conditions. These included pedophilia, voyeurism, public masturbation, compulsive hiring of prostitutes (or sex workers) and use of peep shows, "tendency to commit rape," and unwanted masochistic desires. The researchers reported positive outcomes in a number of cases and concluded that one of the drugs they used "shows promise as a treatment for paraphilias." But complications occurred in each of the twelve cases described: one patient experienced nausea and vomiting; some lost the ability to ejaculate or have an erection altogether; others showed a complete absence of sexual feeling or interest and became severely depressed; and every patient subjected to prolonged treatment suffered bone mineral density loss, putting them at risk for osteoporosis.

Another problem with antiandrogen drugs is that their effect on a person's libido is typically global rather than selective. Imagine that you wanted to reduce only harmful or ill-directed lust—toward a prepubescent child, for example, or an adult who was tempting you away from your well-considered monogamous commitment. You would likely be disappointed. Current biotechnology is not sensitive enough to reliably deliver on these sorts of person-specific goals.

Anti-attraction drugs

Anti-attraction drugs are a bit trickier. For one thing, interventions into the attraction system are less studied than those that primarily affect libido. Some blunt chemical instruments currently exist, but the nature of what makes a partner attractive in the first place is little understood and is likely to be highly variable. Insofar as they could be shown to work, anti-attraction drugs would probably reduce the obsessive thoughts characteristic of early-stage romantic relationships, or the chance that an initial spark of attraction would lead to longer-term attachment.

Whether it would be possible to block attraction to particular individuals or groups remains to be seen. However, research from the late 1800s is suggestive. Around that time, the Finnish anthropologist Edvard Westermarck observed that people living in close proximity during the first years of their lives—brothers and sisters, cousins raised together for arranged marriages, genetically unrelated kids growing up in tight quarters on Israeli kibbutzim—become desensitized to each other as potential sexual partners. Although the exact mechanism underlying the Westermarck effect remains unknown, some scientists think it might involve learning certain olfactory cues.

Whatever the mechanism, there seems to be a critical period for the desensitization to take place, which raises the intriguing possibility that the right treatments (pharmacological or contextual) could reopen this window. This, then, might make it possible to essentially trade one's romantic feelings of attraction to someone for nonromantic feelings that are closer to how people typically feel about siblings or cousins. If that turns out to be possible—and we are completely speculating here—it might not remove companionate feelings but just feelings of sexual desire. In short, the Westermarck effect shows that the brain is at least capable of selective "negative sexual imprinting" (basically tagging someone as *not* a potential mate), and so shutting down romantic feelings for an otherwise eligible partner.

Now for something a bit less speculative. Donatella Marazziti is an Italian neuroscientist based at the University of Pisa. She has been trying to find out whether serotonin might play a role in early-stage romantic attraction. Her hunch came from observing that this all-consuming period, characterized by obsessive thought patterns and nervous preoccupation with the tiniest of details (kind of like our jealousy example earlier), might have more than a passing resemblance to obsessive-compulsive disorder, or OCD, which has already been linked to low levels of serotonin. The idea was that the "same lack of serotonin that results in an OCD patient's believing that touching a door five times upon entering can guarantee safety may also be behind the way you constantly, compulsively think about a new squeeze when you are in the honeymoon phase of a relationship."

As she predicted, participants who had recently fallen in love—still in the intense first stage of a relationship but prior to having sex—had levels of serotonin similar to those of a sample of OCD patients, both groups having lower levels than healthy controls. As Marazziti and her coauthors conclude, "It would suggest that being in love literally induces a state which is not normal." Indeed, retesting the lovers at twelve to eighteen months revealed that serotonin levels had returned to baseline, at which point their "obsessive ideation regarding the partner" had disappeared as well.

Given the findings of Marazziti and colleagues, one possibility is that drugs used to treat OCD could dampen at least the obsessive aspects of a nascent amorous relationship. As it happens, patients with OCD respond most reliably to SSRIs, that same class of antidepressants whose ability to diminish the sex drive is now well known. And as we discussed before, SSRIs can also sometimes lead to "emotional blunting" of higher-level feelings involved in romantic attraction and relationships generally: 80 percent of SSRI-using patients in the study we mentioned "reported less ability to cry, worry, become angry or care about others' feelings."

Again, if you're trying to maintain a relationship, not caring about your partner's feelings is a lousy idea. If you're trying to end a relationship or keep it from developing into something more, however, maybe this type of treatment could grease the wheels.

Anti-attachment drugs

Finally, anti-attachment interventions. There just isn't a whole lot of concrete evidence that existing technologies could completely sever a long-term human pair bond, although breakdowns in attachment obviously happen on their own all the time (just ask anyone who has gotten over their ex). There is, however, compelling evidence for such a possibility in other mammals with analogous mating habits—namely, those voles we met in the chapter on oxytocin (OT). As we saw, injecting OT could foster a pair bond in the prairie voles without any actual mating. And crucially for our purposes in this chapter, this kind of effect could be reversed.

In one study, injecting female prairie voles with either an OT or a dopamine blocker caused them to lose their monogamous tendencies; that is, they failed to show any partner preference as a function of copulation. As Larry Young put it: "They will not bond no matter how many times they mate with a male or how hard he tries to bond. They mate, it feels really good, and they move on if another male comes along." Likewise, pair-bonded male prairie voles injected with a dopamine blocker—at a specific site in the brain known as the nucleus accumbens—failed to show characteristic mate-guarding behaviors and became more receptive to interactions with novel females.

As we mentioned before, most scientists who study human attachment believe that something like the prairie vole bonding mechanism is preserved in mammals over evolutionary time and shows up in our species as well. Nevertheless, we are not aware of any scientists who have injected blockers for OT, vasopressin, or any other neurochemical into the brains of human subjects to find out what happens to their romantic attachments. Not only would this be hard to justify to a university ethics committee; it would also presumably not inspire many volunteers.

There may be a way around this scientific hurdle, however. Consider a recent headline from *VICE* magazine: "How to Bio-Hack Your Brain to Have Sex Without Getting Emotionally Attached." The author, Sirin Kale, writes about an "an all-too-familiar situation for many people: You decide to have sex with someone whose personality you find repugnant, whom you have no interest in dating, only to find yourself bizarrely attached to them in the morning." Now, one thing you could do to avoid this situation is to think twice about going to bed with someone whose personality repulses you. But suppose it's too late for that, and you're already deep in the ardent throes of your dubious decision. Can you somehow immunize your brain against its own tendency to form a sex-based emotional attachment?

According to Larry Young, you can. The trick is to avoid eye contact with your partner during sex. Research shows that prolonged eye contact causes the release of oxytocin in the brain; this, in turn, increases the chance of forming a bond. "When you're having sex with someone," Young explains, "you're making an intimate connection with their face and eyes particularly. This is going into your brain, and it's inherently rewarding. Love and attachment are very much like addiction. They have a lot of the same chemicals. So if you can divert that information from coming in by not having that eye contact, that will help."

Certain illicit drugs may "help" as well. According to Young, cocaine and methamphetamine boost dopamine secretion, which is also involved pair bond formation. If you take a drug to raise your dopamine levels "prior to an intimate moment," he says, "it won't have the same impact later. The specialness of the sex, and the differential caused by the dopamine release won't be so high."

Finally, alcohol can foster attachment-free sex (which might be one of the reasons why they often go together). But the disruptive effect of alcohol on pair bonding seems to be different for males and females. At least, that is how it seems in the case of voles. "When male voles drink alcohol they become promiscuous and it prevents them from bonding," Young says. For female voles, it is exactly the opposite. Alcohol "increases the likelihood they will bond prematurely."

Voles in the wild are not usually big on alcohol. So you can guess that these findings come from a lab. In a crucial experiment, Young and his collaborators allowed male prairie voles to self-administer "substantial amounts" of alcohol (the voles were happy to oblige). Once tipsy, they were introduced to a female prairie vole for purposes of mating. "Normally, if the male vole mated with a female, the next day when we put him in a three-chambered cage containing three female voles, he'll opt to sit with the vole he previously mated with," Young says. This is classic prairie vole monogamous behavior. But if the male was drunk during sex, he won't be as inclined to get cozy with the same mate the following day. Instead, "he'll prefer the novel females."

For female prairie voles, as we noted, the effect of alcohol on sex and bonding is different. When given the choice to huddle with their original mating partner or a tempting stranger, the females that drank alcohol during "cohabitation" showed "a robust and statistically significant preference for the partner over the stranger."

Young and his colleagues argue that the differential effects of alcohol on pair bonding in male versus female prairie voles are "mediated by neural mechanisms regulating pair bond formation and not alcohol's effects on mating, locomotor, or aggressive behaviors." In other words, "alcohol has a direct impact on the neural systems involved in social bonding in a sex-specific manner." Whether this is also true for humans remains to be seen.

Another, more speculative lead for an attachment-dissolving intervention comes from a medical condition known as Capgras's delusion. In this tragic delusion, an individual reports believing that a spouse, sibling, or close friend has been replaced by an impostor who shares identical visual features. Patients suffering from this condition are able to recognize faces, but the automatic emotional connection to familiar faces that most of us experience just doesn't come online. That lack of emotional connection leads to the belief that the person must be an imposter.

One explanation for this phenomenon is that neuroanatomical pathways responsible for responding to familiar visual stimuli have become damaged or degraded. This account fits nicely with the oxytocin-vasopressin-dopamine model of attachment, which as we mentioned involves combining social-identity cues (such as the physical features of a person) with a network of positive emotions.

Anti-attachment interventions of the future, then, might mirror the Capgras syndrome—ideally without inducing its delusive aspects—by interfering with this integration in a targeted way.

The ethics of current use

Taken together, these findings suggest that it may soon be possible to block or diminish lust, attraction, and/or attachment using pharmacological strategies. In fact, it is already possible to achieve some of these effects, if in a blunt and haphazard way. One question this raises, then, is whether these current, messy drugs should ever be used to affect relationships, notwithstanding the risk of various side effects and the fact that most of these uses would be off-label.

As we have been stressing, most drugs prescribed today are intended to treat individual symptoms of acknowledged diseases or disorders, not the ordinary experiences of interpersonal strife that may be, at least to an extent, rooted in an individual's biology. As we saw with the jealous antique shop manager, however, this barrier can in principle be side-stepped by explicitly treating an accepted pathology—OCD, in that man's case—while knowing full well that the real goal of the treatment is to improve the person's romantic relationship.

The ethics of prescribing drugs off-label is tricky. Sometimes the evidence concerning appropriate doses, benefits, and risks has changed since the manufacturer's label was finalized. If you're prescribing a drug for the purpose it was originally intended for, in a way that is consistent with the best available evidence, and the evidence just happens to have changed since the label was printed, hardly anyone would seriously object.

On the other hand, if you're prescribing a drug for a purpose it was *not* originally intended for, the most likely scenario is that there won't be enough high-quality evidence concerning the effects of the drug for that use. You might then be exposing the patient to unknown harms, which weighs against the practice.

This takes us back to an earlier point. The drugs we currently use for individualized symptoms *are already* affecting our relationships, but in ways and under conditions about which we are mostly clueless, because the evidence base consists primarily of case studies and anecdotes. Sure, you can approach treatment for someone dealing with jealousy "as if" they had OCD—and cross your fingers that he stops haranguing his wife. So long as you have a plausible on-label use for the drug that just happens to be associated with a person's relationship difficulties, you are probably good to go. But we should be going about this in a much more systematic, scientific way.

The risk of pathologization

So why aren't we? What's stopping us, in other words, from deliberately researching the interpersonal effects of common medications? One reason, we suspect, is that people are afraid of pathologizing love and relationships. Since doctors are only allowed to prescribe drugs that society regards as medicine, and since medicine is typically used to treat patients who have some kind of disease or disorder, the fear is that prescribing a drug to help someone's intimate relationship would imply that the relationship is pathological—when maybe it's just down to hard times.

This fear is understandable given the current paradigm. But the paradigm needs to change. If you remember our definition from earlier, drugs are just chemicals. You can call them medicine if you want to, but chemicals don't know if you have a disease they're supposed to be treating (so that they can somehow confine themselves to acting only within the definition of that disease). Instead, they just do whatever they do, whether you intended them as a cure for a pathology or just believe they could improve your life.

We should be open to the idea that certain chemicals can increase people's happiness or well-being, on balance and under the right conditions, without our first having to invent a disease for those chemicals to be addressing. In the case of improving relationships, the idea is not to subsume yet another domain of ordinary life under the disease labels of medicine so that people can gain access to certain drugs. Instead, it would be to take certain drugs out of the domain of disease labels and ask whether they could improve certain aspects of relationships.

In other words, rather than medicalizing romantic relationships, perhaps we should consider demedicalizing certain drugs. In the meantime, we should collect as much evidence as we can about the advantages and disadvantages of using chemicals for such nonmedical ends—since, again, they are very likely having effects on those ends whether we intend them to or not.

We need to keep thinking about the ethics. Our goal for the next chapter is to construct a more general ethical framework for handling biotechnology that would block or diminish potentially problematic forms of lust, attraction, or attachment. To do that, we are going to zoom out from the specific drugs and technologies we have surveyed so far, so that our framework will be broad enough to handle even future biotechnologies that will be more potent than the ones we have now.

CHAPTER 10

CHEMICAL BREAKUPS

BONNIE'S PARTNER, ROB, wore big pewter biker rings on every finger. He had an awful temper. After a trifling incident early in their marriage, says Bonnie, "he started smashing me in the face with his knuckles. He grabbed my hands and bent them backward, breaking one of my fingers." As she later told a journalist: "I was in shock. I was stunned. But I didn't leave. A few hours after the incident, Rob broke into tears and told me how sorry he was. I loved him so much I believed him when he said it wouldn't happen again."

Bonnie's experience of what she characterizes as love for her abuser is far from unique. Such cases may even represent a form of Stockholm Syndrome—a condition in which victims form a strong emotional bond with their attacker as a way to cope with ongoing trauma. In these cases the abused partner is reluctant to end the relationship and may actually defend the aggressor when others try to intervene. Harmful relationships can be literally addictive for some individuals, leading them to chase constantly after momentary highs of emotion only to crash back down into inevitable despair.

Needless to say, people who are in physically, verbally, or emotionally abusive relationships do sometimes manage to leave them, but their feeling of visceral attachment to their partner may not abate, even months or years after the breakup. This can result in long-term suffering that interferes with or even precludes close relationships in the future. Take this recent posting from an internet support group:

> About 10 months ago I broke up with my abusive partner, but I am unable to get over him. I have tried to date other men but I always end up breaking things off with them, just wishing I was back with my high school boyfriend. I know he is a terrible person, but I am unable to have other relationships. I'm scared that I'll never be able to be in love again.

A certain amount of pain and difficulty in intimate relationships is unavoidable. Sometimes it is even beneficial, at least instrumentally, because suffering can lead to personal growth and self-discovery. This doesn't mean we should intentionally hurt our partners so that they can grow from whatever it is we're inflicting on them. It's just to acknowledge that hardship, in relationships as in other areas of life, can sometimes strengthen our character if we respond to it in the right way.

But sometimes suffering is truly bad and not constructive to a higher end. And some emotional attachments are simply dangerous. Either they can trap a person in a cycle of violence, as with Bonnie, or they can prevent a person from moving on or forming healthier relationships, as with the woman in the second example. The ancient biological machinery behind love doesn't always function to our advantage in everyday life. It can misfire in many ways and cause, or increase the chance of, serious harm.

Among other things, this machinery can push a person to cheat on a spouse or partner (infidelity); fall in love with someone who will never return their feelings, leading to despair or even suicidal thoughts or behaviors (unrequited love); form the delusion that another person is in love with them, leading to stalking or other harmful behaviors (erotomania); feel angry and violent when love is rejected (spurned love); feel crippling jealousy for no good reason (pathological jealousy); feel uncontrollable sexual attraction toward a child (pedophilia); develop an obsessive admiration for, and devotion to, a dangerous cult leader (cultish love).

This list is obviously not exhaustive. For one thing, it doesn't include examples of relationships, desires, or behaviors which violate certain traditional moral values but not progressive ones. Consider that our ancient biological machinery can also push a person to fall in love with someone outside of their racial group, religion, social station, or caste; fall in love with someone of the same sex or gender; desire consensual swinging, or group sex; form voluntary BDSM relationships; or develop an interest in a harmless sexual fetish.

Most progressive-minded people wouldn't see anything wrong with these relationships, practices, or forms of desire, but many conservative people do—and they might be motivated to stamp them out. What these contrasting lists highlight is a deep disagreement about which models for love and desire are good and which are bad in society at large. Accordingly, there may be very different intuitions about whether it makes sense to try to modify a person's biology to help them fit particular standards or norms. Is there any way to discuss the ethics of such modification without first resolving the wider disagreement?

In order to move forward, we'll proceed in small steps. First, we'll focus on a case that pretty much everyone agrees is troubling, namely a case in which someone has an addictive attachment to a violent abuser, as we have seen with Bonnie. We will try to see if it could be justified to modify someone's biology in a case like that. As always, our point of departure is that no chemicals should be used without accompanying psychosocial interventions. But if the argument for a biological supplement can't succeed in this seemingly most straightforward of situations, it will hardly succeed when things get more complex.

Drugs and abuse

We think everyone can agree with the following: a feeling of very deep attachment to someone who is severely and habitually violent toward one is likely to be harmful overall. Could there be a legitimate role, then, for some kind of anti-attachment drug in such a scenario, possibly to help the person leave the relationship?

Instantly, we hit a problem. What if the abused person doesn't want to leave the relationship? Many people in abusive relationships develop coping strategies where they reformulate the violence in their mind ("My partner wouldn't hit me if they didn't really love me"). Or they come to believe, however misguidedly, that a certain amount of violence or abuse is acceptable, as long as it is offset by other aspects of the relationship that they value to a greater degree.

Who would be the one to decide about applying a drug in such cases? It wouldn't be the abuse victim, who doesn't want to break their feelings of attachment. So it seems it would have to be some other person or group effectively forcing the drug on an unwilling target.

This raises a lot of red flags. There is a deeply troubling history, particularly in psychiatry, of extreme and unjustified forms of paternalism in dealing with people who are considered mentally compromised but do not want the "help" they are being offered. It is a history of coercively applying sometimes dangerous medications to people for their own alleged good. In the case of certain relationships, it is of course possible that the feelings of attachment might be so extreme that they really do make a person "lose their mind." Some degree of paternalism can be justified when an individual lacks decision-making autonomy and poses a threat to themselves or others. So if there were indeed a case of undeniable, serious, and persistent abuse, where a victim claimed that everything was fine and there was no need to worry, there might be an argument for overruling their decision and intervening against their will.

But the risk of *unjustified* paternalism looms large. In general, people should be extremely hesitant to assume that they know what is in somebody else's own best interests, especially when that person is an adult, and even more especially in the matter of close, personal relationships, where it can be very hard to know what is going on in terms of intimate details.

Even in cases where there is a clear argument for using coercive measures to break up a bad relationship—with or without the help of a drug—there will usually be a more appropriate target for the intervention than the person being abused. Namely, the abuser. So while disrupting a person's feeling of attachment to an abuser could be helpful in some cases, the most salient source of harm would be the abuser's violent behavior. The primary goal then should be to stop the violence, not to drug the victim into a state of detachment.

What then about cases where the victim does want to leave? Bonnie reported feeling so much love for her husband that she believed him when he said that the violence "wouldn't happen again." But of course that is not the end of the story:

> Life became hell after that. For the next two months the abuse was nonstop. Rob kept me in a constant state of terror. I'm not a drinker, but he'd toss a rum and Coke in my face and say drink. He'd punch me in the stomach or kick me in the thigh if I didn't. I started walking on tiptoes around him, fearful of everything I'd say and do. He dislocated my shoulder several times. He'd lift me up by the ankles and bang my head against the floorboards in the living room.

With such a brutal person in the house, why didn't Bonnie file for divorce? A part of her wanted to leave, she said, but "another part of me hesitated. He'd cry and show such remorse that I'd forget my own pain. He'd become romantic and sweet, and I'd fall in love with him all over again."

From the way Bonnie describes her feelings here and throughout the rest of the interview, it becomes clear that she knows she needs to leave the relationship. She has a rational, second-order desire to leave it, but her more visceral feeling of romantic attachment is standing in the way. Her ancient biological machinery, in other words, is badly misfiring and causing her to feel emotionally addicted to someone who beats her up. She is in conflict with herself, and she wants a resolution.

Before we say more, we want to clarify a few things. First, many people who are in abusive relationships seem to believe they cannot leave them, not because they have some kind of emotional attachment to their abuser but because they are financially or otherwise economically dependent on their partner. They may also be afraid of putting their children in danger by leaving. We don't know how many cases there are of people who actually know they need to leave a bad relationship and who have the means to leave it, but who fail to do so because of their own unshakable feelings of attachment. We won't speculate about the percentages. But we want to make clear that those are the cases we are concerned with here.

The second thing we need to clarify is that insofar as Bonnie's partner, Rob, is physically abusing her, he is doing something wrong and criminal. So it is he who is obligated to make a change, not Bonnie. We do not wish to blame the victim. To make sure there is no confusion, then, let us impose some further assumptions. We are going to assume that Bonnie (a) has access to an effective anti-attachment drug, (b) desires to use it to extinguish her feelings for Rob, (c) reasonably believes this will help her pursue her own higher-level rational goal (namely, to leave the relationship), and (d) would take the drug voluntarily.

If the use of a pharmacological cure could ever be morally justified in the context of a real-life relationship, it seems to us that this modified "Bonnie" case would be among the most promising:

1. The feelings in question (namely, those of attachment to someone it is bad to be attached to) are clearly undesirable, both objectively and from the perspective of the person experiencing them; and
2. The person wants to use biotechnology, believing reasonably that it will aid in the achievement of a higher-level rational goal; and this would be done voluntarily, under conditions of informed consent.

Other promising cases would share this basic structure. For example, individuals who are desperately, and unrequitedly, in love, who have no prospect whatsoever of having their feelings returned, and who are therefore descending into utter despair (we have actually received e-mails from people in this situation, asking how they can get an anti-love drug); individuals who fall unwillingly (yet irresistibly) for someone other than their monogamous partner; individuals who have uncontrollable, and unwanted, sexual desire for young children. And so on.

Although the case of pedophiles who hate their own desires may not elicit the kind of sympathy typically felt for victims of intimate-partner violence, it might actually be an even clearer instance of when the use of a drug would be justified. Because of the stigmatization, social exclusion, self-loathing, and guilt that often go along with pedophilic urges, some people with pedophilia consider suicide for altruistic reasons, to avoid harming children or degrading themselves. So here, and in the modified Bonnie case, it seems that some kind of intervention would be permissible.

Even so, you might want to argue that there is a crucial distinction between simply trying to diminish the lust, attraction, or attachment—which probably would be justified in these cases—and using a biochemical intervention, specifically, to do it. We will look at that argument next.

Means to an end

What is behind the move to draw such a distinction? One argument goes like this: a person suffering from unwanted and genuinely harmful attachment or desire should not turn to a drug to address the problem but should instead rely on more traditional, nonbiomedical methods—psychotherapy or counseling, for example. We raised this issue when we introduced you to Sofia, the woman who contacted Brian asking for an anti-love drug so she could leave her husband.

But we don't need such dramatic cases to see the wisdom of this approach. Just take the more mundane example of trying to get over your ex. You might try all sorts of things that don't involve pharmacology. You could dwell on all the ways your ex has really hurt you. You could avoid your ex when you see them—and stay away from the places they tend to be. You could delete all of their e-mails, block their phone number, and stop looking at their pictures on Facebook.

These are all sensible courses of action. We agree that these kinds of methods should be preferred—or at least tried first—for several reasons. First, they are probably safer and would have fewer unpredictable side effects than most drug-based interventions for the foreseeable future. And they would probably best preserve a person's ability to reflect on the deeper issues that led to the need for a breakup in the first place.

As we mentioned, the pain of heartbreak often does have an upside. It forces us to slow down, reflect on what happened, and try to figure out how to avoid getting into a similarly bad relationship in the future. As the bioethicist Neil McArthur has written: "A skin contusion or broken bone can heal just as well without the pain associated with it. But we feel pain over a broken relationship as we work through what went wrong and begin to conceptualize a life that is different from the one we expected to have."

Learning to make the pain go away, he writes, "is precisely what it is to heal." But if we avoid this process by turning to medication, "we risk trading short-term pain for a more lasting, and less treatable, unhappiness."

There is certainly a risk of turning to medicine to solve your problems when what you need to do is learn and mature. But not all medicine is created equal. As we mentioned in our discussion of MDMA, some drugs may actually increase your ability to reflect more deeply about the reality of your situation. During our interview with Ben Sessa, the British psychiatrist and expert on MDMA, he brought up some of these considerations. "I think, in a way," he said,

> that MDMA provides an opportunity for self-reflection, which is an enlightening experience, which you can then use to either leave a relationship or bolster a relationship. To suggest that MDMA is a love drug that always brings people together would be wrong. It

> can bring them together during the session; it can bring them together to reflect on the fact that they shouldn't be together in the relationship. I think it's good that it provides that opportunity.

Still, it is important to be cautious about adding drugs into the mix of any relationship. Starting with more traditional methods to get out of a bad relationship is exactly the right idea. Yet in some cases, traditional methods do not work. Maybe a person doesn't have the strength or willpower to tackle the bigger-picture issues sufficiently to break things off without the help of a chemical ally. Suppose we add a third item to our list of best-case-scenario ethical constraints. In some cases,

3. The person cannot overcome the undesirable feelings without the help of biotechnology.

This third condition, together with our first two, would seem to create the strongest possible moral justification for the use of an anti-love drug of some sort. First, it would be clear that the feeling was undesirable, objectively and subjectively, which would rule out contentious cases like same-sex love. Second, the point of using the drug would be to promote a person's higher-order goals or commitments, and this would be done voluntarily. And third, the drug would be necessary to achieve those goals: traditional methods alone would have been exhausted or thoroughly considered and ruled out.

Establishing strict boundaries

These three boundary conditions set a very high bar for the ethical use of chemical "cures" for any given situation. Have we set the bar too high? You might think that the last constraint—the necessity requirement—is too stringent. Consider: What if it were possible to diminish some destructive form of lust, attraction, or attachment without using a drug, but just much harder? What if the emotional suffering involved in using traditional methods was so severe and protracted that the instrumental value of that pain (for personal growth, self-discovery, and avoiding bad relationships in the future) seemed truly dubious? What if it seemed dubious even if you couldn't rule out the possibility that some deeper life insight would be gained by sticking it out?

People will disagree. In the academic debate about chemical enhancement, at least two main camps are always warring it out. On one side are "bioconservatives," who tend to be resistant to technological changes that will significantly affect the human condition. They would probably remind us here that even great and seemingly unbearable suffering can impart important lessons and that people should hesitate to use drugs to solve their problems. "With suffering comes understanding," they might say.

"Bioliberals" are on the other side, and they tend to be more open to technological change. They might point out that traditional methods of altering our feelings also affect brain chemistry, only less directly than many drugs and often less effectively. "Sometimes suffering is just suffering," they would say, arguing that individuals must ultimately decide what's best for them.

Referring to such arguments, the American bioethicist Erik Parens has written that if pharmacological and traditional methods "can both achieve the same end—improving how one experiences herself and the world—then [some would say] it is irrational and perhaps inhumane to prefer the more strenuous and expensive means." That is, it could be considered "irrational *not* to take a shortcut when improving human well-being is the destination." Summarizing the view of bioliberals (which for the record is probably not how he would identify himself), he writes: "We should be slower to imagine that suffering leads to growth and understanding, and quicker to remember that sometimes it just crushes human souls."

No one would deny that there can be great value in experiencing the world as it really is—in its heartbreak and agony as much as in its multitude of joys. But we should resist a naïve realism which equates the way the world "is" with how it happens to appear to us during normal waking consciousness (that is, without the use of any drugs). Earlier we used the metaphor of a magnifying glass. We said there is a sense in which looking through a lens distorts what we see, because it presents things to us differently than how they appear in our normal vision. But when we look through a magnifying glass, what we see is no less real for being refracted through a lens. In fact, we can see more real things with a magnifying glass at our disposal than we can without it. Similarly, while there might be some drugs that hide things from us, by covering our emotions and dulling our pain, other drugs may take the cover off. That was the whole point of our chapter on MDMA and psychedelics.

But even if some drugs do introduce a true distortion, our duty to experience the world "as it is" is not absolute. Sometimes relieving suffering can trump the value of appreciating reality, or the indeterminate probability of growing as a person. Besides, bioconservatives might not be giving enough consideration to what economists call "opportunity cost." Extreme suffering can sometimes help inspire self-development, but similar insights may be achieved without the agony. What's more, too much suffering may prevent a person from achieving *different* insights or pursuing projects that might be just as valuable or even more valuable, despite the relative lack of pain.

As a general rule, we suspect the person whose life is most directly affected by the bad relationship—or the process of going through a breakup—will be best placed to identify the point at which their pain and suffering is more than they can bear. No universal line can be drawn in the sands of human suffering. People do have to decide for themselves.

Let's update our third constraint on the permissible use of chemical cures, then, relaxing the necessity requirement to allow for individual judgments about where to draw the line:

3. The person cannot overcome the undesirable feelings without the help of biotechnology, or at least cannot do so without incurring extraordinary psychological or other costs that the person reasonably judges to be unacceptable, all things considered.

Here, then, is a summary of all three conditions we've proposed:

1. The feelings in question are clearly undesirable, both objectively and from the perspective of the person experiencing them;
2. The person wants to use biotechnology, believing reasonably that it will aid in the achievement of a higher-level rational goal; and this would be done voluntarily, under conditions of informed consent; and
3. The person cannot overcome the undesirable feelings without the help of biotechnology, or at least cannot do so without incurring extraordinary psychological or other costs that the person reasonably judges to be unacceptable, all things considered.

Thinking ahead

As with any biotechnology, a chemical cure for lust, attraction, or attachment may bring about benefit or harm. By thinking through the ethical factors involved in cases like the ones we have considered, our hope is to nudge things in the direction of benefit. But this is only the beginning of the conversation. In the real world, not all situations will resemble the ones that we have argued are maximally promising. When the relevant drugs are developed and come to market, as McArthur notes,

> policymakers, doctors, and individuals will all have to make judgments about the value of such drugs in various kinds of real-world situations. For most of us, our experiences of falling in and out of love are central, formative events in our lives. In the face of a revolutionary technology that allows us to manipulate this process, an ethical stance becomes urgent and compulsory, and a neutral position impossible. Will public funds aid the development and improvement of the drugs? If they become available, will they be designated prescription-only or sold over the counter? Will public and private health plans cover their cost? When should doctors prescribe them? How will public health officials educate the public about them? What would we tell a friend or loved one who was debating whether to take them?

These are precisely the sorts of concrete questions society should be asking itself. Not everyone will agree with our ethical framework, and that is all right. If there is a better one, we would genuinely like to see it. But we do need to acknowledge, as a society, that however wonderful it can feel to be in love—or even just think we are in love—this most central of human emotions can also be dangerous. We have argued here that when it is dangerous, we may sometimes have good reasons to try to escape its powerful thrall—even if this means swallowing a pill.

CHAPTER 11

AVOIDING DISASTER

SO FAR WE HAVE TRIED TO SHOW that love drugs and anti-love drugs are not some made-up possibility for the future: biotechnologies are currently available that can have an enhancing or degrading effect on the neurochemical bonds that underlie romantic love, and these could plausibly be used to help maintain some good relationships and end some bad ones. Drugs and other technologies with even more powerful effects on relationships will likely continue to be developed, and we have suggested that the time is now to set up an ethical framework for handling this situation.

By looking at specific cases and thinking through likely benefits, risks, and contextual factors, we argued that some potential uses of love drugs or anti-love drugs would be ethically appropriate. But we have mostly left untouched the bigger-picture questions about implications for society if people started using biotechnology in those ways.

What this means is that even if you happen to agree with our take on the specific cases, you might still think that society as a whole would be better off without any type of love drugs in the first place, since their very existence might cause more problems than they help to solve. Now, as we argued, it's really too late for this kind of response, since love-altering drugs are already available and some are in widespread use. The question then becomes one of regulation—who should get to use the drugs that do exist? Under what conditions should they be made available? How should we respond to the development of new relationship-affecting biotechnologies? Should we try to prevent their coming into existence or restrict their use? And so on.

In answering these questions, let's remember that any new technology poses risks. This is true whether it's a relationship drug, the internet, or mobile phones. The mere possibility that a technology could have a downside is never a sufficient reason to reject it, however alarming the potential downside might be. Instead, the harms that might come with the (mis)use of the technology have to be weighed against the benefits that could come with its responsible use, considered in light of plausibly achievable laws, policies, and practices that could alter the balance for the better.

In the case of love drugs, we have argued that some couples may find them helpful in restoring their relationship, if that is what they judge to be best, especially if there are vulnerable children involved who would otherwise be harmed by their separation. In the case of anti-love drugs, we have argued that the ability to sever emotional ties with an abusive partner could benefit some people, so long as social and legal routes were not neglected. Other benefits could include the voluntary reduction of pedophilic desires, unwanted adulterous impulses, or overpowering feelings of unrequited love.

Still there are serious risks to consider on a larger scale. One potential issue is that by bringing love and relationships into the domain of medicine, no matter how successful we might be in alleviating suffering or promoting interpersonal well-being, we would lose something of great value in our intimate lives. This is sometimes expressed as a concern about "medicalizing" ordinary human experiences. As it turns out, the fear of medicalization can be broken down into a number of more specific concerns—for example, concerns about reducing complex problems to apparently simple technical fixes—and we will look at each of those in turn in the final chapter. Here we want to address a different but related worry, which is that interventions into the biological side of love could be used to try to homogenize the sexual and relational landscape.

The basic idea is that some communities might try to discourage or eliminate sexual preferences or forms of love or desire that are not actually wrong or harmful—or that might even be good and positively worth pursuing—based on unjust stigmas or misguided values. If we are going to get on board with neurotechnologies that can be used to refashion human love and sexuality, it will be necessary to double down on efforts to dismantle intolerant social forces, while developing supportive policies to make sure the technologies are not misused. Gay, lesbian, bisexual, and other sexual orientation minorities, as well as those who subscribe to nondominant relationship models like polyamory, would be especially vulnerable to coercive treatment. This is a real concern and we need to think deeply about how to address it.

As a starting point, we should rule out all involuntary use of love drugs or anti-love drugs. Just as it is illegal to spike someone's drink at a party, it should be a crime to administer love drugs or anti-love drugs to any person under any condition without their informed consent. This means that children, who are incapable of providing consent to such interventions, should be strictly protected, and this must be a priority for law and policy. But even adults may face profound social pressure to change how they experience or express their feelings of love and sexual desire, so that merely having the option to change might place an unfair burden on them. In essence, they would be forced to justify why they decided to "retain" their sexual orientation or relational disposition, when joining the majority was a real possibility. Clearly, what it means in practice to give informed consent without undue coercion cannot be analyzed in a cultural vacuum. Let us consider some of these issues in detail, starting with the historical context.

The bio-oppression of sexual minorities

The misuse of biological interventions into love and sexuality is not hypothetical. Nor is it a forgotten wrong from ages past. The practice of so-called conversion therapy in the United States—designed to "cure" gay and lesbian individuals of their sexual and romantic feelings—carried on until at least the 1970s with the full-throated endorsement of the mainstream mental health profession. And as late as 2012, a U.S. federal judge ruled that such "therapy" cannot be outlawed, even when it is conducted on minors, since it constitutes a protected form of religious speech. It is still being performed in at least some fundamentalist religious communities in the United States and elsewhere to this day.

Historical efforts to modify the sexual desires of those with predominantly (or exclusively) same-sex attractions have included applying electric shocks to the hands or genitals, pairing nausea-inducing drugs with the presentation of homoerotic stimuli, reconditioning impulses to masturbate, deep brain stimulation, psychoanalytic therapy, "spiritual" interventions including peer pressure and prayer, and even brain surgery.

Some of these more invasive approaches are unlikely to be tried today, but others persist. Earlier in the book we mentioned yeshiva students. Psychiatric drugs are being given to Orthodox yeshiva students in Israel at the request of religious leaders and marriage counselors as a way of suppressing same-sex sexual feelings, so that the "patients" will find it easier to comply with rigid norms forbidding homosexual behavior.

In the United States in 2015, the Obama administration argued that such sexual orientation change efforts (SOCE) can have "potentially devastating effects" on the lives of "gay, lesbian, bisexual and queer youth," and that they are therefore neither medically nor ethically appropriate. The Human Rights Campaign made a similar argument, adding that SOCE fail to do what they purport to do—that is, change one's sexual orientation.

The available evidence supports these claims. Specifically, the physical, mental, and medical risks associated with common approaches to SOCE, though hard to estimate with any precision, can be extremely severe, up to and including death by suicide when they do occur. And apart from a smattering of nongeneralizable, anecdotal reports to the contrary, such approaches overwhelmingly do not seem to work.

But is this a good strategy for arguing against SOCE? We believe it is not. The main reason we oppose such a strategy is that it ties human rights law and policy to a technological situation that may be only temporary. Yes, current SOCE are largely harmful and ineffective—but what if future SOCE actually did work and could be made relatively safe? In other words, what if current techniques were improved upon (in a manner of speaking) so that they were in fact successful at changing sexual orientation, and with an acceptable benefit-to-risk profile as judged by those who sought them out? Should SOCE then be legally permitted, whether on minors or anyone else? Could such efforts be morally permissible, whether legal or not?

These are the main questions we want to address in this chapter. They are worth taking seriously. Based on current trends in research, it is not implausible that in the not-too-distant future, scientists will know enough about the genetic, epigenetic, neurochemical, and other brain-level factors involved in shaping sexual orientation that those variables could be effectively—and we will just stipulate, safely—modified with the application of biotechnology. Likely combined with existing or perhaps refined psychobehavioral approaches, it might then be possible for individuals with same-sex sexual attraction to redirect their erotic desires to the opposite end of the spectrum (roughly, convert from a homosexual to a heterosexual orientation), as well as presumably the other way around (that is, people with a heterosexual orientation could convert to a homosexual one). Let's call such potential future technology *high-tech conversion therapy*, or HCT.

Two notes of clarification before we go on. First, on terminology: we will sometimes use the labels homosexual and heterosexual in this chapter as we have just done, drawing on the common "folk" taxonomy of sexual orientation—which also includes bisexual—despite the fact that there is, in our view, a great deal more to sexual orientation than these labels suggest. (For example, what is the sexual orientation of an intersex or genderqueer person who does not fit neatly into the male-female dichotomy? What would it even mean for such a person to have a same-sex or opposite-sex attraction?)

Second, we do not support the creation or the use of HCT. Instead, we want to show that the very prospect that it might be developed is a threat to the ongoing human and civil rights movement for gay people and others with minority sexual orientations. Currently, this movement is centered on a "born this way" response to unjust discrimination—essentially, an appeal to the innateness of sexual orientation—as well as a related response that draws on the notion of unchangeability (as captured in the song lyric, "I can't change, even if I tried," from the chorus of "Same Love" by Macklemore & Ryan Lewis). Together these responses are commonly thought to support an argument that goes like this:

Premise 1. It is wrong to discriminate against someone on the basis of an unchosen, unchangeable characteristic.

Supporting analogy. For example, it is wrong to discriminate against someone on the basis of their race or sex, which are unchosen and unchangeable characteristics.

Premise 2. Sexual orientation is an unchosen and unchangeable characteristic (like race or sex).

Conclusion. It is wrong to discriminate against someone on the basis of their sexual orientation.

Is this a good argument? No. And it may actually backfire on gay rights advocates in the long run, as we'll explain. To start with, the supporting analogy is breaking down as transitions in identity based on sex and gender are becoming more widely recognized—recall our discussion of transgender individuals from a while back—and now even "transracial" identities are being discussed by some scholars. Does being born with a certain trait or disposition automatically entail that it cannot be changed? No. Is it okay to discriminate against someone because of their sex, gender, or race, even if those traits can in fact be changed? No. So you can see potential problems right away. But even if the argument were sound, which it is not, our point is that it would be vulnerable to empirical refutation insofar as HCT is possible.

We don't take the "born this way" argument lightly. It likely has been effective in improving attitudes and behavior toward people with a minority sexual orientation. As our friend, the musician and scholar Mark Bailey, wrote to us in response to some of our earlier work, it is important to see this real-life context:

> The timeline of events in history that led to [the] "it's not a choice" counter-argument clearly shows that this is not inherently a matter of gay-rights activism, but, rather, a necessary grasping unto something presented by a segment of the scientific community that . . . could enable a needed moment of relief from relentless attacks against the soul. "It's not a choice" has been a way to survive.
>
> In other words, many have had to endure endless abuse for their non-heterosexuality: daily bullying, loss of employment, public humiliation, discrimination, excommunication, loss of family and friends—sometimes murder or suicide. And then, finally, in the face of all this, struggling gay men and women had something to say that would cause some attackers to pause for a second by virtue of a few magic words: it's not a choice. There are people who reclaimed their lives because of those magic words.
>
> Whether, in the end, sexuality is truly a choice doesn't matter. But if someone who has endured malice all his life for feelings he felt no control over, finally, can get up in the morning and face the world with confidence while others back off, just a little, then maybe we can better understand this particular timeline of progress and not mistake it for irresponsible activism.

Mark is right. The "born this way" argument captures something very real, and very meaningful, about the experience of most sexual orientation minorities. Just like people with heterosexual desires, people with nonheterosexual desires do not just wake up one day and decide to be attracted to members of one sex category or another. Instead, the desires are rooted deep inside them, typically from the time they first become aware of having sexual desires at all. And just like people with heterosexual desires, people with nonheterosexual desires typically feel that the desires cannot be substantially changed, no matter how much it might make their lives go better in a homophobic world.

Yet a longer-term strategy is needed. In our view, securing fundamental rights for individuals with minority sexual orientations is just too important a project, morally speaking, to be based on faulty premises or left hostage to biotechnological fortune. It might be possible to ban the development or use of HCT, but counting on that possibility seems risky as well. We think it would be better, all things considered, to argue that whether or not sexual orientation can be chosen or changed is completely irrelevant to how we should treat people with minority sexual orientations—socially, legally, or otherwise. People should be supported in forming consensual romantic relationships, if that is what they want, whatever their sex or gender and whatever the sex or gender of the people they love and desire. To discriminate against people because of their sexual orientation is wrong whether it is innate or immutable or not.

Which is to say, as others have said before us, gay rights are human rights. All of us deserve to live, love, work, play, and flourish in our families and communities, and to build lives connected with others in accordance with our values—at home, at work, and in public—without fearing for our safety or survival. And that is true whatever our sex, gender, race, religion, disability status, or sexual orientation happens to be, independent of the degree to which we can exert some influence on these aspects of our identities or ways of life.

To set these claims up properly, we need to start by drawing some distinctions that often get lost in the public debate. What does it actually mean to change one's sexual orientation? Is this the same as choosing to be gay, straight, bisexual, or something else? And why does it ultimately matter?

Identity and orientation

Can you choose to be gay? Most campaigners for gay rights would say no. But in January 2012, former *Sex and the City* star (and more recently, gubernatorial candidate) Cynthia Nixon, who identifies as gay, said that she chose that option from among alternatives. In an interview, she put it like this:

> I gave a speech recently, an empowerment speech to a gay audience, and it included the line "I've been straight and I've been gay, and gay is better." And they tried to get me to change it, because they said it implies that homosexuality can be a choice. And for me, it is a choice. I understand that for many people it's not, but for me it's a choice, and you don't get to define my gayness for me.

She went on to say, "A certain section of our community is very concerned that it not be seen as a choice, because if it's a choice, then we could opt out. [But] why can't it be a choice? Why is that any less legitimate? It seems we're just ceding this point to bigots who are demanding it."

What were some members of the gay community worried about? The concern was based on Premise 1 of the argument we gave above: the idea that if people can't choose whether they are gay, straight, or bisexual, then it seems unfair to discriminate against them on those grounds. If too many members of the public were to take Nixon's statements about choosing to be gay seriously, then, the fear was that this would damage one of the most important conceptual planks in the contemporary platform in favor of gay rights.

Yet already we are mired in confusion. To see why, we need to draw our first distinction: between a sexual orientation-based *identity* and a *sexual orientation* full stop. "Being gay," we suggest, is the former: it's a personal identity category that is usually based in a sexual orientation (namely, having primarily or exclusively same-sex sexual attraction). But the precise mapping between a person's identity (in this sense) and their sexual orientation is not always clear-cut: people often have wiggle room in how they make sense of their inner lives, including the sexual aspects, and in how they communicate this self-understanding to others. With few exceptions, each individual is their own authority on how this mapping plays out. In other words, if you choose to identify as gay, then you are gay. No one can get inside your head and tell you otherwise.

Now, if you find yourself overwhelmingly attracted to members of the opposite sex, and not at all to members of the same sex (we want to reiterate that these are oversimplifications; neither sex nor gender are simple binaries), then you would be a bad citizen of your language community to go on and apply the label "gay" to yourself. You would be bound to cause confusion. We do not, as a rule, get to make up our own new personal meanings for words and expect others to play along.

But if you are capable of feeling attraction to members of more than one conventional sex or gender category, as many people are, and if you prefer to orient your romantic and sexual behavior around the same end of the spectrum as yourself, then go ahead and consider yourself gay. Who you are in this biographical, identity-label sense is not a fact of nature. It's more a form of shorthand you might use to help organize your self-conception, or simplify a process of self-disclosure with others. It can also be a way of consolidating insights about what is most central to your experience of yourself, to your desires, your relationships, and your way of moving through the world. And it can even be a way of signaling your core beliefs, ideological commitments, or allegiance to a community of which you feel most strongly a part—along the lines of explicitly political identities like Democrat or Republican. In other words, an identity claim like "I am gay" is really a placeholder for a longer conversation.

The question of who you happen to be sexually attracted to across time and circumstance—in terms of the sex and gender characteristics of those you find yourself most irresistibly drawn to—is less up for debate and is largely a different question. This is what we have characterized as your sexual orientation, "full stop." As of right now, most people do not have significant control over their sexual orientation in terms of their essential erotic desires. In other words, in contrast to our sexual orientation-based *identities* (gay, straight, bisexual, pansexual, sapiosexual, or what have you)—over which we do have some measure of control—our more basic sexual orientations are, as far as scientists can tell, not really up to us. Instead, they are largely given to us by our genes, exposure to certain hormones in the womb, and other factors that are not under our command. (As the mayor of South Bend, Indiana, and 2019 U.S. Democratic presidential candidate Pete Buttigieg said in a campaign speech, using religious language to make a similar point, "If my [sexual orientation] was a choice, it was a choice that was made far, far above my pay grade. . . . If you got a problem. . . your problem is not with me—your quarrel, sir, is with my creator.")

So what should we conclude? For some people, there certainly is room for choice with respect to their sexual orientation-related identity. Cynthia Nixon is one such person, and she chooses to be—or to identify as—gay. But even Cynthia Nixon, if she is like most people, cannot just choose her basic sexual desires, that is, her sexual orientation (full stop).

Perhaps the "born this way" argument can be rescued, then. Sure, you might be able to decide how you want to identify, or act, on the basis of your sexual orientation, but if your sexual orientation itself is largely determined by biological factors outside of your control, then isn't it still unfair to discriminate against you simply for *having* nonheterosexual desires?

Yes, it is. Not only unfair but morally vacuous. There is nothing wrong with having *or* acting on nonheterosexual desires, so long as the behavior is consensual. But the issue of choice remains a red herring. To see this clearly, consider the perspective of bioethicists Tia Powell and Edward Stein. Using the term LGB to refer to lesbian, gay, and bisexual individuals, they write:

> Even if sexual orientation is not chosen, most of what is legally and ethically relevant about being an LGB person *is* the result of conscious choice. Actually engaging in sexual acts with a person of the same sex, publicly or privately identifying as an LGB person, and marrying a person of the same sex and raising children together are *choices*. In other words, an LGB person could decide to be celibate,

> closeted, single, and childless. Support for LGB rights is precisely support to make these *choices* and to do so without fear of discrimination or violence. The right simply to have same-sex attractions, without the right to act on these desires or to express the related identities, would be an empty right indeed.

This is a vital point. Rights for LGB people and other sexual orientation minorities should not be limited to those dimensions of their lives or identities that are in fact immutable, even if that category does currently include their sexual orientation. So the gay rights movement should move away from lack of choice as a major plank in its arguments and activism. In addition to the reasons given by Powell and Stein touching on choices in behavior, there is also the possibility that sexual orientation itself could one day be subject to choice through the use of HCT.

Assuming that happened—that HCT was in fact developed—would the gay rights movement suddenly concede that minority sexual orientations are problematic? That such orientations should in fact be changed, as currently argued by the religious right? That it would be permissible to discriminate against individuals with such orientations if they decided not to convert?

If it would not make these concessions, then choice is not the issue. Whatever is the issue, then—presumably the fact that neither having nor acting on a minority sexual orientation is inherently wrong, and that individuals have a moral right to have consensual sexual interactions and form romantic relationships with whomsoever they choose—should be the focus of activist efforts going forward.

This takes us right back to Nixon and her "bigots"—those people who want to deny equal rights to nonheterosexual couples. To insist that sexual orientation is not a choice, she says, would be to allow the religious right to "define the terms of the debate." Her point is that resting the argument for gay rights on the issue of choice implies there might indeed be something wrong with nonheterosexual desire and that it is only the impossibility of changing such desire that makes it acceptable. But surely that isn't what advocates for gay rights really think.

There is an obvious analogy to drive this point home. As Dan Savage—of monogamish fame—writes: "Religious conservatives go on TV, and knock on doors, distribute pamphlets, proselytize, and evangelize all over the country in an effort to get people to do what? To change their religions. To choose a different faith."

In other words, "Faith—religious belief—is not an immutable characteristic. You can change your faith. And yet religious belief is covered by civil rights laws and anti-discrimination statutes." The only time you hear "that a trait has to be immutable in order to qualify for civil rights protections," he writes, is when conservatives talk about homosexuality.

Savage is right. Few human traits are "legally salient and yet cannot be changed." For example, there are currently procedures that make sex change possible (on some conceptions of sex), but this element of choice does not mean there should be a different legal standard for discrimination on the basis of sex. And if HCT is ever developed, the same should be true for legal standards bearing on sexual orientation.

Ethical implications

Our discussion so far has been pretty abstract. We want to zoom in on a specific situation now, where a person might actually be motivated to change their sexual orientation in real life. We will focus on the case of the yeshiva students we mentioned earlier. Consider this statement by Professor Omer Bonne, director of the psychiatry department at Hadassah University Hospital in Jerusalem:

> Some behaviors put Haredim [Orthodox Jews] in conflict with their values and cause them mental problems, even to the point of depression. . . . My view concerning drug treatment in such cases has changed. For example, when I was young, idealistic and less experienced, whenever I had a case of homosexuality [or] masturbation . . . I would say: "Homosexuality is not a mental problem, masturbation is certainly not a mental problem or even a medical problem. I do not treat people who do not have a medical problem." Over the years, I saw that people who do these "awful" things suffer terribly

> because of the conflicts they create. Those urges, impulses or behaviors place them in conflict with their society, and then they become depressed. In these cases, I would indeed prescribe medicines that block these conditions.

What should we make of Professor Bonne's remarks? In the first place, we want to highlight that the drugs in question—selective serotonin reuptake inhibitors, or SSRIs—really can block a person's libido, making them lose their sexual desire. We talked about this effect in earlier chapters. This isn't quite the same as actually reorienting that desire (for example, from homosexual to heterosexual), but it shows the power of biological interventions.

Second, we believe that what ultimately needs changing in such situations are the background religious norms that stigmatize non-heterosexual feelings and behaviors—not the feelings or behaviors themselves. As the psychologist Douglas Haldeman has rightly argued, members of his profession must attempt to "reverse prejudice" rather than sexual orientation:

> Homophobic attitudes have been institutionalized in nearly every aspect of our social structure, from government and the military to our educational systems and organized religions. For gay men and lesbians who have identified with the dominant group, the desire to be like others and to be accepted socially is so strong that heterosexual relating becomes more than an act of sex or love. It becomes a symbol of freedom from prejudice and social devaluation.

It is no surprise then that many gay people have turned to—or been pressured into—conversion therapy as the only way to live out their lives with some semblance of happiness. For those choosing to be true to their identity and sexual orientation rather than continue living a lie to keep the peace, it is not easy to just pack up and leave, though many have done so. At the same time, in many places around the world gay people (and other sexual orientation minorities) face life-threatening violence and even murder simply for being who they are. In those cases, even if they are willing to break ties with their family or religion, they cannot do so without risking death. This is unjust beyond description, and human decency demands that such prejudice be contested and finally changed.

The need to fight institutionalized homophobia, religious fundamentalism, and Bronze Age attitudes about human sexuality cannot be overstated. But that is a long-term project. What can be done in the meantime? To turn back to our case study, how should the very real, present-day suffering of those religious students be addressed, given that the repressive norms of their insular communities are unlikely to liberalize any time soon? Could the use of sexuality-altering biotechnologies, high-tech or otherwise, in such cases ever be morally justified?

In answering this, we are simply assuming that the students are adults, because we have already stipulated that children should be protected from HCT and other such biotechnologies. But this doesn't make the ethical analysis any easier. You might argue that adults should be free to intervene in their own biology to pursue whatever they take to be their highest values, even if they don't have an underlying medical problem for which the intervention would be an appropriate treatment. Indeed, this is exactly the approach of some transgender theorists who argue that access to medical technologies, such as hormones and surgeries, should be given to those who seek to manifest their gender identities (in part) through relevant changes to their bodies, without this resting on the notion that "being trans" is in any way an illness or psychopathology.

We agree that transgender identities are not pathological and that access to surgeries or hormones for gender affirmation should typically be made available to those whose lives would be improved by the bodily changes they seek. Primarily, this is because such procedures tend to allow the person to inhabit their body and live out their identity in a way that is most authentic for them, increasing their overall well-being. Yet we would still be inclined to support access to gender-affirming technology (GAT) in cases where a person's dissatisfaction with their unmodified body was, to whatever extent, related to the existence of unjust social pressures in the environment, such as the narrow-minded policing of sex or gender norms and associated stigmas, mistreatment, or violence.

That was basically our argument earlier in the book. There we tried to show that even if the social script for love—or sex, or gender, or sexual orientation—has serious problems and should be changed, this does not automatically mean that people are unjustified in seeking to change their biology in cases of persistent tension. Indeed, in response to some of our earlier work on HCT, the philosophers Candice Delmas and Sean Aas argued that if HCT were in fact available and requested by an adult with a same-sex sexual orientation—even if their desire to convert was motivated in whole or in part by the restrictive norms of their religious community—"clinicians would often be permitted, and sometimes even required, to prescribe reorientation to patients who suffer from their sexual orientation, given the medical principles of beneficence and respect for patient self-determination."

It is a difficult conclusion to swallow. At least it is for progressive bioethicists like ourselves who simultaneously believe the following two things: (1) there is nothing wrong with having a same-sex sexual orientation (so no one should ever want to use HCT in the first place), and (2) there is nothing wrong with having a transgender identity (so no one should object to the use of GAT for someone who desires it, if it would make their life significantly better).

In an ideal world, of course, there would be no homophobia or transphobia and people would seek whatever biological interventions they desired—if any at all—as a means of pursuing their highest values through autonomous self-expression. But we do not live in an ideal world. And for this reason, Delmas and Aas actually do not support the development of HCT. (They do not say what their position is on GAT.)

In fact, Delmas and Aas have argued that while HCT might very well benefit individuals, both now and in the future, it would be better overall if the technology were not made available, given our nonideal world. In this world, where nonheterosexuality is stigmatized and heterosexuality valorized, they warn that simply having the option to convert would be harmful to sexual minorities on the whole. This harm, they suggest, could manifest in three main ways: (1) pressures to convert, (2) expectations on sexual minorities to justify their non-heterosexual orientations, and (3) conversion being understood as the rational course of action for those who fall outside the norm.

Consider first the likely pressure to convert. While some pressure might arise out of hostility or a lack of tolerance for sexual nonconformers, Delmas and Aas argue that it could also arise out of genuine concern for the well-being of sexual minorities. If only they would join the winning team and become heterosexual, some might say, their lives would go so much better (as indeed they might). To make the choice easier, and the pressure more intense, family members, clergy, or even the government could offer to pay for HCT. As a result of these and other pressures, the option to convert might become an offer that is very difficult to refuse. According to Delmas and Aas, such a situation would place a severe and unjust burden on sexual orientation minorities.

Part of this burden would be an increase in the perceived need to justify one's decision not to convert—the second harm raised by Delmas and Aas. Just as bisexual people are sometimes expected to explain why they pursue same-sex couplings even though heterosexual romance is a real possibility for them, the existence of HCT could make it so that even those who have no heterosexual desires "owe" society an explanation for their deviance. Conversely, those who start off with heterosexual desires, or who acquire them, might not be expected to explain their decision to stay or become that way, as this would be seen as the right and sensible choice.

Indeed, under such oppressive conditions, conversion might widely be seen as the rational choice for those who are not heterosexual by birth or by choice—harm number three for Delmas and Aas. In fact, if you prioritize welfare in your theory of rationality, it might *be* the rational choice. But what is rational for individuals within a group can still be socially harmful if it promotes greater intolerance or injustice toward the group at large. In practice, this harm could befall anyone who maintained a nonheterosexual orientation despite having the option to convert—and despite the growing number of "success" stories from those who did change their orientation and found greater acceptance in the wider community. In this way, sexual orientation minorities tempted by the promise of a better life could find it difficult to resist becoming complicit in their own oppression, or even the erasure of their "kind" of person. The upshot would be a powerful reinforcement of heteronormative domination.

These three harms of HCT, according to Delmas and Aas, "dwarf the potential benefits to individuals that could be gleaned if the technology were available." Taken together, they think, the harms make a "compelling case for the idea that sexual orientation should remain outside the individual's control, and therefore that [HCT] ought not to be created."

Whether or not you find this analysis convincing, it does raise important concerns. It also leaves open important questions. One of these questions is what to say about people who have primarily or exclusively heterosexual attractions who might want to change their sexual orientation to homosexual. This is not a hypothetical situation.

According to the English writer and radical feminist Julie Bindel, in the late 1970s a group of lesbians in Leeds, calling themselves revolutionary feminists, made a controversial proposal that "resonated loudly" for herself and many other women. They began calling for all feminists to embrace lesbianism as a matter of political necessity and philosophical purity. The idea, roughly put, was that men as a class were the enemy—agents of an oppressive heteropatriarchy—and one should, literally, not sleep with the enemy.

The movement reached its height in 1981 with the publication of a booklet, "Love Your Enemy? The Debate between Heterosexual Feminism and Political Lesbianism." In this work, the revolutionary feminists argued that women who slept with men could not be feminists. The book did not insist that women must sleep with women, but for women who were not sexually attracted to other women, this version of feminism implied a life of celibacy. In theory HCT would offer a solution to women who found themselves tempted by political lesbianism but were only sexually attracted to men.

Might conversion from heterosexuality to homosexuality as a way of supporting a morally or politically motivated gay or lesbian identity be permissible, given the objections to such conversion posed by Delmas and Aas? One reason to think it would be permissible is that people with heterosexual orientations are not in fact sexual minorities. Thus, it could be argued, there would be less social pressure for them to change their unchosen, innermost desires, which would eliminate some of the concerns about structural coercion raised by Delmas and Aas.

But this would create a troubling situation: members of a sexual majority would be permitted to use HCT to modify their sexual orientation as a way of conforming their first-order desires to their higher-order preferences or values; whereas current sexual minorities, some of whom might sincerely wish to change their own sexual orientation for principled moral, political, or philosophical reasons, would not be permitted to do so, thereby being forced to submit to a constraint that did not apply to those in the sexual majority.

Seemingly, then, either everyone should be permitted to use HCT, so long as they are mature, fully informed adults acting on their most considered values, or no one should be permitted to use it. Given the current situation in which relatively few people with heterosexual orientations have expressed a desire to become sexual minorities (even in a political sense), while a great many sexual minorities face pressure from their respective communities to convert, the safest vote might be for no one. This would suggest that we should try to prevent HCT from coming into existence, as Delmas and Aas propose, or else ban it even for adults if it does become available—something that could in principle be accomplished by passing the right legislation.

Current technologies

We will not try to settle that issue here. Instead, we will turn our attention to existing interventions, like the ones actually being used by Haredim in our case study (namely, antidepressant medication). As Professor Bonne made clear, these students are in fact suffering from depression. The ultimate source of that depression is a mismatch between their biology—their same-sex attraction or desire to masturbate—and their restrictive social script, which includes their own presumably sincerely held religious beliefs. We have already argued that in such cases it is the social script that should ultimately change. But how does one actually do that? You can't wave a magic wand to make it change; it takes time and political activism, and in the meantime the students are depressed. Even if one thinks they should not have access to HCT for the reasons we have already covered, should they also be barred from "low-tech" drugs, like SSRIs? Remember, the SSRIs are primarily intended to lower the students' libido, but they may also at least partially alleviate their mismatch-borne depression.

This sets us up for a dilemma. Either we can help the individual and at the same time strengthen the objectionable background norms, or we can resist the norms by refusing to help the individual. This dilemma may even extend through time, creating further complications. In other words, while prescribing the drug might very well help an individual in the here and now suffer less severely, the norm reinforcement might disadvantage future generations of sexual minorities in the same community. We might think that for the sake of the long-term battle against regressive norms we should allow individuals to suffer in the here and now in order to help weaken and perhaps end those norms, to the hopeful advantage of future individuals who would get to exist under more tolerant conditions.

We see no easy way out of this bind. One place to look is at feminist critiques of cosmetic surgery considered as a kind of enhancement technology, which offers a (very) rough parallel. The idea here is that women living in societies that overvalue youthful bodily characteristics can achieve greater confidence through surgeries that aim to fit them into society's mold for feminine beauty. But this makes women and surgeons complicit in propping up the norms that are largely behind these body image problems in the first place. Instead of turning to surgery, therefore, people should fight against the norms of physical appearance that are so restrictive as to cause unnecessary suffering.

The earlier problem arises here as well. Anyone accused of being complicit in unjust norms by performing cosmetic surgery must choose between helping individuals while supporting oppression, and letting people suffer to take a stand against those norms.

Feminist philosophers such as Margaret Olivia Little have tried to address such dilemmas. The painful reality of individual suffering in the here and now (even if that suffering is only due to unjust social pressures) presents a genuine problem in need of a solution. How much personal well-being in today's imperfect world must be sacrificed on the altar of future societal progress in changing problematic norms? Ultimately, Little splits the difference between these two considerations, recommending that medical professionals try to strike a balance: between protesting the unjust norms (and refusing to profit from them) and simultaneously helping their patients comply with them to alleviate suffering.

In practice this could mean doing the surgeries pro bono—or charging only a minimal fee—so as not to profit from unjust norms, while at the same time actively fighting against the norms by speaking out, presenting patients with alternative, nonsurgical options designed to boost self-esteem, writing critical op-eds in newspapers, or donating to campaigns aimed at promoting more sensible conceptions of beauty.

Douglas Haldeman has given a similar argument with respect to conversion therapy for religious individuals:

> Ideally, the individual ultimately integrates sexual orientation and spirituality into the overall concept of identity by resolving antigay stigma internalized from negative experiences in family, social, educational, and/or vocational contexts. But what of the individual

> who, after careful examination of the aforementioned factors, still feels committed to an exploration of changing sexual orientation or of managing sexual identity? Even with data to prove that all who request a change of sexual orientation are acting out of internalized social pressure, we would be hard-pressed to deny such individuals the treatment . . . they seek.

What this discussion highlights, at base, is that ethical dilemmas concerning emerging biotechnologies cannot be resolved in an academic vacuum. To the contrary, a much wider debate is taking place in society over what sorts of values we should hold in the first place with respect to things like love, sex, and relationships (and nearly everything else as well). And plainly this broader conversation—between the insights of progressivism and the insights of conservatism, as well as between the forces of secularism and the forces of religion—will continue to shape the moral ends toward which human beings collectively and individually strive, regardless of what technology is actually in hand and regardless of what pontificating bioethicists may argue in their books and papers. At the most fundamental level, the relevant question—what we have called the *basic technology / value* question—becomes, how can we use new technologies for good rather than ill, while simultaneously trying to reach a functional consensus on what sorts of things actually are good or ill in the first place? Progressive-minded people clearly have their work cut out for them in this overarching project.

CHAPTER 12

CHOOSING LOVE

IF YOU HAVE MADE IT THIS FAR, you are probably at least open to the idea of love drugs playing a role in our society. But you might also worry that something special about love would be lost in the process. Part of the magic of love, it seems, is precisely that it is so mysterious—that it can take us over completely, as though by a force outside ourselves. Do we really want to put it under a microscope, much less douse it with a bunch of chemicals from a lab?

As Wordsworth wrote in the "The Tables Turned," "Our meddling intellect / Mis-shapes the beauteous forms of things." Perhaps we have to "murder" love, to borrow his famous phrase, if we are going to "dissect" it. And if that is the price of developing love drugs or making them more widely available—well, it might be better to just call the whole thing off.

Carrie Jenkins, the philosopher who introduced us to the dual-nature view of love, argues that we should resist these kinds of intuitions. Trying to understand and even influence how love works, she thinks, is not some abstract quest for intellectual stimulation. Instead, *not* knowing how love works, whether along its biological or psychosocial dimension, makes us "deeply vulnerable." This is because the majority of us base some of the most important decisions of our lives on whether we at least believe we are in love. As Jenkins argues, treating love as profoundly important yet entirely incomprehensible "shouldn't strike us as normal."

We think she's right. And we are skeptical about the idea that the value of love resides in its sheer mysteriousness—in our inability to understand it—rather than how it acts in our lives. Neither unpredictability nor ignorance alone makes something worthwhile. And even if we did have a complete scientific understanding of love, we would not be able to predict exactly who we would fall in love with or how the relationship would play out. Adding love drugs to the mix won't change that. As we said earlier, we are dealing in nudges and probabilities in this book, not chemical determinism. So perhaps there is enough mystery after all?

Here is an analogy. Imagine you are eating a piece of chocolate cake (or whatever flavor you like best). You fork off a bite, put it in your mouth, and let the deliciousness start to unfold on your tongue. The way that feels is wonderful. It has significant value to you—maybe too much value if you're like us and one piece of cake is never enough. But now imagine that you helped make the cake: you and your partner, say, spent the last few hours in the kitchen together, preparing the ingredients, measuring them out, mixing them up, and putting it all in the oven. Does the cake taste any less delicious to you now? Does knowing the recipe, the chemical makeup of the various ingredients, somehow rob your tongue of the flavor it so craves?

Love is not the same as chocolate cake. There are a million important differences, we know. The point of this example is that knowing how something works does not preclude the possibility that you may experience the full value of that thing. Studying love and relationships scientifically, then, and using the fruits of that investigation to try to make them better, does not entail that love loses all its magic. It is possible that some couples might find even more to value in love when they take a closer took, because they could experience it in novel ways or explore it together from a whole new perspective.

Reading an account of love that draws on evolutionary theory, for example, might inspire a couple to feel connected to their distant ancestors. They might see it as meaningful that those ancestors would, presumably, have experienced similar romantic feelings of their own, by virtue of an ancient brew of neuropeptides and neurotransmitters, shared across a great expanse of time. They might go on to reflect on the role of love (so conceived) in the perpetuation of the species, and feel a part of something larger than themselves. Or they might feel an affinity with nonhuman animals that form analogous bonds, and imagine what "love" must be like for them.

Or they might not. Who knows. Different couples will feel differently about love and about the various ways it can be better understood. Some couples will avoid a scientific approach altogether and will have no desire to apply a drug to their relationship. Still others may have valid concerns about bringing love into the domain of medicine and medical technology, not just in their own case but for anyone. According to the sociologist John Evans, many people "have reached the normative conclusion that they do not want to live in a world where increasing swaths of human experience are under the logic of medicine. There are, or should be, experiences that use an older logic, which are under the jurisdiction of another profession or under no jurisdiction at all." Significantly, he concludes with this warning: "We can all fear the medicalization of love."

Facing up to medicalization

What is behind this powerful intuition? In our research, we have identified four main worries that people tend to have about medicalization, and we cannot finish this book without addressing them head-on. We'll start by looking at the worries at a more general level and then see how they apply to the case of love.

Worry 1: The Pathologization of Everything

Medicalization can transform ordinary human differences and experiences into pathologies, redefining human diversity as so many shades of illness through an ever-expanding application of disease categories and labels.

Worry 2: The Expansion of Medical Social Control

Medicalization can expand the influence of the medical establishment, raising anxieties reminiscent of *One Flew Over the Cuckoo's Nest* about people in white lab coats coldly enforcing what they think is normal. It can also create openings for pharmaceutical companies and other medical capitalists to sell us drugs we don't need for diseases we don't have (or that have simply been invented out of whole cloth), thereby expanding the power of Big Pharma to meddle in our lives.

Worry 3: The Narrow Focus on Individuals Rather Than Social Context

Medicalization can reframe social problems as individual shortcomings, taking resources and attention away from the wider contextual factors that may be creating the "need" for treatment in the first place.

Worry 4: The Narrow Focus on the Biological (or Neurochemical) Rather Than the Psychological

Medicalization can lead to a sort of bioreductionism, favoring simple understandings of phenomena that are in reality much more complex. This can lead to an undue emphasis on molecular-level interventions, promoting a picture of humans as so many bunches of matter to be rearranged, as opposed to conscious subjects who respond to reasons.

These are serious concerns. There is no doubt that medicalization can have bad consequences—or perhaps be bad in and of itself. But it can also have good consequences, so we have to be careful. Unwanted pregnancy, for example, is not a disease, but many people regard the medicalization of abortion as a positive development, as it meant that abortion was no longer considered a criminal act and it could be done more safely.

There are many other examples of medicalization that one might think have been on balance beneficial, such as the medicalization of epilepsy, formerly a "spiritual" condition; the medicalization of alcohol addiction, formerly "deviant drinking"; and the medicalization, and hence treatment, of pain. Though medicalization has an ominous association with the empowerment of medical professionals at the expense of everyone else, the truth is more complex. Many medical technologies have allowed people to gain more control over their own lives in a way that has genuinely helped them flourish. Not all medicalization is bad.

But not all medicalization is good, either. What about the medicalization of relationships and love? We can go through each of the worries that have been raised and see how they apply in this domain.

Worries 1 and 2: The Pathologization of Everything, and the Expansion of Medical Social Control

One way to understand the concern about pathologization is to refer back to the history of medicalization as applied to sex and sexual attraction, as we touched on earlier. As critics have long noted, the subjugation of normal, natural sexual variation to the disease labels of medicine has been extremely damaging; there are countless examples of harmless sexual and relational differences being overtly and inappropriately pathologized throughout the course of prior centuries.

This worry is not confined to the past. If tampering with people's relationship orientations, for example, becomes more acceptable, we can imagine that some people's desire for, say, polyamory might be framed as something that needs to be treated. And we can see many people wanting to turn to biotechnology to help them conform to society's prevailing (monogamous) expectations. We explored that exact scenario in an earlier chapter. Going beyond this, we can imagine the invention of various supposed relationship disorders, perhaps promoted by pharmaceutical companies in their hungry pursuit of profit—things like "commitment phobia" or perhaps "adultery proneness syndrome." Large companies could lead the push by marketing "normality-enhancing" drugs to people who were previously content with their sexualities or relationship preferences, or to people who have concerns that are better addressed through more conventional means.

This is not a fantastical scenario. Appreciating the force of this concern, we want to share a recent real-life example of something called "hypoactive sexual desire disorder" (HSDD). This "disorder" was conjured up by members of the drug industry to create a market for a new drug, Addyi, the so-called female Viagra. We will walk you through what happened in some detail because it is a very good example of how pathologization actually works in contemporary medicine.

Inventing a disorder

Hypoactive sexual desire disorder appears to be an instance of what is called condition branding, where companies coin conditions to fit drugs they are in the process of developing. Pharmaceutical companies have to pass regulatory tests before marketing their new drugs, but they can market new diseases whenever they like. They do this by letting clinicians know (often years in advance) that a new disorder is coming down the pipeline. Sometimes companies develop a drug for one use, find that it fails in this area, and then devise a new disease that it could be said to treat. Perhaps what was originally a side effect could become a drug's intended purpose. In the case of Addyi (flibanserin), the drug company Boehringer Ingelheim originally meant for it to be an antidepressant, but when it proved disappointing as a mood improver, it became a treatment for HSDD.

This rebranding was a failure as well, as the FDA advisory committee determined that flibanserin did not particularly increase libido and had too many adverse effects. Boehringer Ingelheim sold this dud to Sprout Pharmaceuticals. When the FDA prohibited Sprout Pharmaceuticals from marketing the drug, Sprout went on a public relations tear. It convinced women's groups, female congressional representatives, and some clinicians to be mouthpieces for its claim that the FDA was sexist for approving drugs that addressed male sexual dysfunction while blocking Sprout's drug, which allegedly addressed female sexual dysfunction. Next Sprout apparently recruited physicians and researchers to give paid promotional talks and continuing medical education courses.

These marketing messages tend to follow a kind of script. The pharmaceutical company's shiny new medical condition has been ignored by the medical community despite its prevalence and its significant impact on quality of life, while other treatments either do not exist, are only partial solutions, or are fraught with devastating flaws. Consistent with this approach, continuing medical education messaging about HSDD stressed that it was a widespread, underdiagnosed, and urgent problem that had profound effects on quality of life, and yet there existed few means of addressing it.

How do you know if you have HSDD? Well, you fill out a diagnostic questionnaire—one that was designed by the pharmaceutical companies. As it turns out, the diagnostic tools for HSDD were largely or solely devised by manufacturers who were developing drugs to address the condition. One questionnaire asks, "Over the past 4 weeks, how often did you feel sexual desire or interest?" The respondent is given options ranging from "almost never or never" to "almost always or always."

There is no established cutoff for what counts as a "normal" score on this question. However, those who felt sexual desire for about 50 percent of their waking hours would only have scored a 3 out of 5. Only those who felt sexual desire for most or all of their waking hours would have received a 5.

The unsubtle message is that if you do not feel sexual desire pretty much constantly, there may be something wrong with you. Of course, depending on the drug being marketed, constant sexual desire could itself be presented as a dangerous aberration. You can see how this sort of thing works.

This is not to deny that pharmacological interventions for a perceived low libido—assuming they worked—could be beneficial for some women who feel that their lives are missing something because their interest in sex is not as strong as it could be. The problem is that drug companies have an active interest in making more women feel like something is wrong with them for not always being interested in sex, or even that they have a medical condition in need of treatment.

We have no doubt that the advent of love drugs—in addition to the ones that already exist—will be seized on by pharmaceutical companies as an opportunity to create new "conditions" to go with them, for which they hope to establish a large market. This is a real concern and there is no getting around it, except to change the way that pharmaceuticals are regulated. At the same time, pharmaceutical companies will seize this opportunity for any new drugs: the problem is not unique to drugs that affect relationships. What is needed is a large-scale, coordinated social movement to resist this kind of insidious disease mongering: through media campaigns, peer education, and putting pressure on legislatures, just to start.

Creating hope

There is some hope. First, the drugs we have given the most attention to in this book—namely, oxytocin and MDMA—are already in existence and would not themselves be a patentable source of profit for pharmaceutical companies (although processes for producing them in large quantities could be). And second, with MDMA especially, insofar as the relationship problems are due to an underlying trauma in one or both partners, the effects may be such that other medications are no longer needed. This is what we saw with the Iraq War veteran, Jonathan Lubecky.

Ben Sessa, the British psychiatrist we met earlier, draws an analogy between mental health issues and having a bad cold. Current pharmacological approaches in psychiatry, he argues, are like ibuprofen and paracetamol: they lower your temperature, painting over the symptoms of the fever, but they don't "kill the bug." He told us that this is why he often sees psychiatry as a "palliative care profession."

"If I meet someone in their twenties who is presenting with severe depression or anxiety or PTSD as a result of child abuse," he told us, "there's a pretty good chance they're going back to see their psychiatrist in their sixties. That's very poor." Extending the analogy, he sees MDMA as an antibiotic that allows us to kill the bug. "So you see people on MDMA therapy coming off their SSRIs that they've been on for years, because they just don't have those symptoms anymore." If that sort of thing could be made to happen on a wide scale, the thinking goes, we would have less of a need for other drug-based treatments.

Another source of hope comes from looking at historical trends. Overall, the disease-oriented focus of medicine seems to be diminishing over time rather than increasing. The older model, which grounded a paternalistic physician-patient relationship in its pretense to objectivity and its clinical-pathologic number crunching, is starting to fade. Emerging in its place is an approach that centers patients' quality of life around their own subjective measures of well-being and general preferences, which lends itself to a more equal physician-patient relationship. In addition, the power of pharmaceutical companies and the medical-industrial complex to conjure diseases out of thin air is being met with increasing public skepticism and awareness—the widespread critical response to Addyi in the popular media is one example.

This does not mean that opportunities for abuse do not persist. But it does show that problematic consequences associated with medicalization—including rampant pathologization of normal, natural differences whose supposed harmfulness is based in flawed moral thinking—are not unavoidable and need not be left unchecked. Social movements, grassroots efforts, and patient advocacy groups have growing leverage in gaining medical recognition for conditions or diagnoses that are consistent with their norms and values, as well as demedicalization of those that are inconsistent with those values—as illustrated by the case of homosexuality—even in the face of resistance from the medical establishment.

The ethicist Kristina Gupta has outlined several measures that could be put in place to diminish the restrictive normalizing pressures of any future technologies. We reproduce these measures here in full as we believe they are important and should be highlighted for serious consideration and advocacy.

1. Include education about sexual diversity (including BDSM, asexuality, and polyamory) in medical and mental health curricula.
2. Institute practice guidelines requiring professionals who prescribe these treatments to provide their patients with information about sexual diversity and referrals to appropriate communities.
3. Require drug companies to undertake qualitative evaluations of the effects of these drugs, and require the Food and Drug Administration to consider this research in decision-making so that drug treatments are evaluated according to more holistic criteria.
4. If drug companies wish to advertise a drug directly to consumers, they should be required to spend an equal amount of money advertising all of the other medically approved treatment options for that particular condition. (Or better yet, direct-to-consumer advertising of such drugs could be banned.)
5. Governments should increase the allocation of research funds for sociological and anthropological research on sexual issues.
6. The Centers for Disease Control and Prevention or other government organizations should undertake public education campaigns designed to educate the public about sexual diversity.
7. Comprehensive sexual education should be required in all public schools, and the sex education curriculum should include information about sexual diversity.

Another way to fight unnecessary pathologization is to rethink the relationship between "drugs" and "medicine" in contemporary societies. According to the current paradigm, for you to have access to certain drugs, they must either be legally tolerated recreational substances (such as alcohol) or be considered medicine. The problem, as we've said, is that medicine is typically seen as being conceptually tied to the treatment of a disease or disorder. This means that if there is a drug that might help you or your relationship in some way—not because there is some objective pathology, but because you happen to be dealing with a problem the drug could help to resolve—you must first receive a medical diagnosis. That way, the drug can be conceived of as medicine and your access to it seen as legitimate (and possibly reimbursed by your insurance).

This whole paradigm, as we have argued, rests on a confusion. At the level of molecules, there is no distinction between "drugs" and "medicine." Drugs are just chemicals, and they do whatever they do regardless of the constellation of concepts or labels that may be swirling above them. Recognizing this, ethicists are increasingly arguing that we should allow the administration of psychotropic substances on the basis of their ability to improve people's quality of life, even when there is no particular illness to diagnose (in the sense of a "mechanical breakdown" of some brain function, for example).

This would defuse the potential problem of the "pathologization of everything" because it would separate treatment (applying a medical technology) from pathology (identifying a disease). Such a paradigm would obviously blur the distinction between treatment and enhancement, a topic of much debate in the recent bioethics literature. For welfare-oriented enhancement theorists, including ourselves, who argue that the goal in either case should be to improve well-being, this would be a welcome shift.

The old specter of pervasive medical surveillance—of oppressive normalization and top-down control—is not going to disappear on its own. With conscious and deliberate effort, however, activists, academics, politicians, and others can do their best to bring about its demise. The takeaway here is that the solution does not lie in avoiding potentially problematic technologies altogether. As long as one does not subscribe to a strong technological determinism, which says that the mere existence or availability of a given technology inevitably produces certain social outcomes, the trick is to think through any problems the technology might create, and then work to prepare the context (social, legal, and political) in which it would be used.

Escaping reductionism

Now let's look at the second pair of worries.

Worries 3 and 4: The Narrow Focus on Individuals and Biology, Rather Than on Social Context and Psychology

What about the concern that medicalization obscures the important communal aspects of well-being? For example, by focusing completely on the neurobiological features of some condition, it may be viewed as a genetic or biological problem and thus treated predominately with drugs—while the social environments that feed the condition are not changed.

Framing complex psychological phenomena as being really just about chemicals in the brain is indeed a common mistake. The case of HSDD provides a nice illustration of this. Here is a passage from a BBC article about HSDD entitled "Libido Problems: Brain Not Mind." See if you can spot the conceptual confusions:

> In recent years, a diagnosis of "hypoactive sexual desire disorder" (HSDD) in women has become more accepted by science. However, there remains controversy about whether the term can or should be used to describe a lack of sexual desire, which may be caused by a variety of psychological, emotional and physical factors.
>
> The latest study [on the question] highlights differences in mental processing in women who have low sex drives. Its author, Dr. Michael Diamond, said it suggested that HSDD was a genuine physical problem.
>
> He recruited 19 women who had been diagnosed with the condition, and compared their brain responses with those of seven others using a functional magnetic resonance imaging scanner, which can measure levels of activation in different parts of the brain by detecting increased blood flow. The women were asked to watch a screen for half an hour, with everyday television programs interspersed with erotic videos.

> In the seven women who did not have the HSDD diagnosis, increased activity in the insular cortices [areas of the brain involved in emotional processing] could be seen. The same did not happen in the women with HSDD.
>
> Dr. Diamond said: "Us being able to identify physiological changes, to me provides significant evidence that it is a true disorder as opposed to a societal construct."

Dr. Diamond is simply mistaken. When he suggests that his experiment provides evidence that HSDD is a "genuine physical problem," he confuses correlation with causation. In a very basic and almost trivial sense, every mental state has a physical explanation; mental events do not occur independently from the brain. To think otherwise would be to subscribe to a kind of Cartesian dualism, according to which our minds are made up of spirit stuff, floating around the vicinity of our heads. Unless you hold a view like that, it would be impossible for the brains of two women experiencing different mental states (such as sexual desires, but also thoughts, motivations, sensations, or any other mental states) to be exactly alike. The point is that these differences in brain activity tell us nothing about what actually caused the differences in brain activity, or whether those differences are pathological, or bad.

To put this another way, the brains of women with low sex drives must be different from the brains of women with high sex drives, because they have different sex drives! Similarly, if John is bored and Mary is not, their brains will have different activity. If we had a scanner with a high-enough resolution to record their brain activity, we would see that the bored brain, compared to the not-bored brain, showed a different physical signature. But this does not show that boredom is a "true disorder."

Of course, this doesn't mean that brain causation isn't real or that we can never detect it. Changes in the brain can genuinely cause other changes in the brain, and so on to changes in the mind. For example, a brain tumor may cause an increase in pedophilic sexual desire, as happened in the infamous case of a forty-year-old schoolteacher (although his story was widely reported, his name was not released to the media). This is a real (and indeed rare) case of "pedophilia: brain, not mind." But in virtually all cases of complex sexual behavior—or any other complex mental phenomenon—we have very little idea about the actual causal chain involved. At best, we'll know just a few steps along the way.

Speaking generally, what causes brain-level differences between people with different mental states or experiences could be genetic, neurochemical, environmental, social, or some combination of the above. The only way to figure out causation is to run an experiment where you actually change or manipulate one of those variables in one group of people and compare their resulting brain activity to a similar group of people who've been left alone—keeping everything else the same.

In this book, for example, we've been writing about the possibility of administering synthetic love drugs to couples undergoing counseling, that is, actually intervening at the level of the brain in an attempt to enhance the couple's mental states, including their subjective experiences of being in love. But then the couple could enhance their own mental states behaviorally—by having sex, for example—which would trigger the release of natural love drugs like oxytocin created by the brain. This just goes to show that there are many ways to effect changes in the brain, some of which might be indistinguishable from each other on a brain scan.

Even if some problem does turn out to be in some sense more "in the brain" than "in the mind," this doesn't tell us what the best mode of intervention will be. Social and environmental stimuli modify brain activity just as pharmaceuticals do. As we've seen, sexual intercourse and orgasm lead to higher levels of serotonin; but so does Prozac. So even if the cause is biological, it remains an open question whether the best treatment is biological, psychological, or social. Indeed, even if the cause is primarily psychological or social, the best intervention may still be biological. The mode or type of intervention does not matter as much from a moral perspective as whether it improves the well-being of the people involved without harming others or violating other moral constraints like respect for justice.

Another facet of the reductionist problem can be seen in a story from the *New Yorker*. It describes a psychiatrist who, all too eager to prescribe antidepressants to his patients, failed to "distinguish between suffering rooted in [their] dysfunctional bodies and suffering rooted in their minds or social contexts." The punch line comes when he asks one of his patients how her antidepressant medication is working. "It's working great," she says. "I feel so much better. But I'm still married to the same alcoholic son of a bitch. It's just now he's tolerable."

A similar point could be made about the domestic abuse example from earlier in the book, where the victim felt an emotional attachment to her abuser. As the philosophers Diana Aurenque and Christopher McDougall argued in response to our initial paper on this topic, "The first and most obviously justified intervention in [this] case is . . . not to drug the victim into unfeeling, but to alert the authorities to the violence and refer her to supportive social and legal services." Similarly, the feminist bioethicist Laura Purdy has argued that administering medicine to women in oppressive social situations may be no better than putting on a superficial Band-Aid: it could lead to far worse outcomes than would be achieved by changing the conditions in which they live.

We agree. Interventions into the social and psychological rather than the biological side of love will very often be the best strategy overall. But as Purdy herself points out, although political solutions may often be better than medicalization for protecting vulnerable people's health and well-being, it does not follow that interventions should never be medical. For even in the best of circumstances some people will need the help of medicine in addition to political change, or to cope with such change. And when that is the case, the medicine should be available.

The key in our view is for societies, through their policymakers, to consider medical interventions as complements to social and political change, rather than as replacements. As Kristina Gupta writes with respect to romance-related technologies in particular, individual-biological and social-structural factors interact with each other in important ways. She writes:

> Interventions aimed at the individual may be effective and may have reverberating effects on the broader social issues, and vice versa. I would simply encourage scholars considering the ethics of biotechnological interventions to address problems with an individual and social component to emphasize the importance of integrating these individual interventions with social interventions and to consider how the two might work in tandem to achieve change. Combined with efforts to address the social factors that contribute to [problematic relationships or forms or states of love] and with measures in place to mitigate the normalizing potential of these interventions, [such] technologies may indeed increase human flourishing.

Choosing love

We would like to finish with a little story. In 2015 the popular *New York Times* column "Modern Love" ran an essay by an academic named Mandy Len Catron, whose experiment with love grabbed the attention of the world. In "To Fall in Love with Anyone, Do This," Catron recounted how over drinks at a bar she and an attractive university acquaintance decided to try an experiment designed by psychologist Arthur Aron. The point of the experiment was to see if it was possible to get people to fall in love in a lab. Here is how she explained the study to her acquaintance:

> A heterosexual man and woman enter the lab through separate doors. They sit face to face and answer a series of increasingly personal questions. Then they stare silently into each other's eyes for four minutes. The most tantalizing detail: Six months later, two participants were married. They invited the entire lab to the ceremony.

As Catron acknowledges, she and her acquaintance were different from the study's participants in several key respects: they knew each other, for one, and they were already on what could be construed as a date. But they followed the rules nonetheless, posing the questions to each other in turn. The now-famous "36 questions" start innocuously, with low-stakes inquiries like who you might want as a dinner guest, and grow increasingly intimate and revelatory: "Tell your partner something you like about them already." After posing each other the questions, Catron and her acquaintance stared into each other's eyes, an experience that was, she wrote, as thrilling and frightening as hanging from a rock face by a rope. In the end, they fell in love.

Catron's article went viral to a degree that shocked her, as she later said in a TED Talk. The article was published on a Friday evening and by Sunday she had been invited to the *Today Show* and *Good Morning America*. The piece got more than eight million views and was translated into Korean and Chinese. Total strangers wrote, e-mailed, and called to ask if she and her boyfriend were still together. Catron said she understood that everyone wanted to know if the study really worked to produce lasting love, but that she had no easy answers. "I want the happy ending implied by the title to my article, which is, incidentally, the only part of the article that I didn't actually write," she admitted. "But what I have instead is the chance to make the choice to love someone, and the hope that he will choose to love me back, and it is terrifying."

That is the message about love we want to leave you with. The idea that love—if you let it, however terrifying it may seem at first—can be an act of will. A decision. A *choice*. Once we see that love is something that we can strive to make happen, or change or enhance, we can turn to the question of means. Asking questions and staring into each other's eyes might do the trick for some. Adding love drugs might be necessary for others. Either way, the agency of the actors will play a central role.

In his 1956 masterpiece *The Art of Loving*, Erich Fromm argues that there is a hidden danger in the popular view that love is something that just happens to you (when you meet the right person, for example) rather than something for which you must take personal responsibility, and work on, and try to improve. As we quoted in the book's epigraph, for Fromm love "is a decision, it is a judgment, it is a promise. If love were only a feeling, there would be no basis for the promise to love each other forever."

Many people are delighted to be swept off their feet in the early stages of a romantic relationship, and just as devastated, later on, when those ebullient feelings start to fade, seemingly out of nowhere and outside of their control. They might even misattribute their changing feelings to something wrong in their partner (or the relationship) and go rushing into a breakup or divorce.

But what if the problem is not so much in their partner or the relationship, but at least partly in their concept of love? What if to love is to practice an art, as Fromm argued, which requires conscious effort and discipline, as well as knowledge and therefore understanding? What if knowing how love works, in other words, right down to the chemicals between us, could help us be better at being in love?

EPILOGUE

PHARMACOPEIA

MODERN PSYCHIATRY IS CHARACTERIZED BY what one drug researcher has called a "wide pharmacopeia of medications, from antidepressants to mood stabilizers to antipsychotics and hypnotics, willingly marketed to doctors by a profit-driven pharmaceutical industry." Do we really need more drugs? We actually think the answer is no. What we need are changes to society: political action that puts human welfare ahead of special interests; resources to help people make good choices about forming and maintaining close relationships; less stress, and more time with friends and family. But so long as we use drugs for medicine—as societies have always done and will continue to do indefinitely—we will need better drugs. More effective drugs. Drugs with milder side effects, with less risk of dependency and abuse, and with the capacity to encourage more serious engagement with the underlying problems that plague our minds and relationships.

The researcher we just mentioned is, once again, Ben Sessa, the psychiatrist we interviewed about MDMA. He argues that the current "pharmacopeia" often does little more than mask the symptoms of our mental and emotional ailments, "leaving the trauma grumbling away beneath the surface." In contrast, MDMA-assisted psychotherapy, and similar regimes involving psychedelic agents like psilocybin, appear to do a better job of targeting the root of the problem, making it easier for patients to address their trauma. This could be something from childhood, from a past relationship, or maybe from a current one.

The greater apparent root-level effectiveness of MDMA and certain psychedelics, compared to most current medications, is one reason why Sessa believes it is so important to take the clinical route in testing these powerful drugs, with the aim of getting government approval and solid regulations in place. On that firmer ground we could avoid a kind of unchecked "chemical utopia"—that is, a world with little or no regulation, where people self-medicate with MDMA or psychedelics, taking them wherever and however they like.

Sessa sees this government-approved, data-driven route as the only plausible way forward. "I care passionately about psychedelics," he said,

> Psychedelics are the newest thing we have in pharmacology and psychiatry for the past seventy-five years. Everything else is old hat compared to this. I really want to see them available to everybody. And we need to do that the very boring, slow way that we're doing it. There's no alternative. Somebody asked me at a conference, "Do you worry about the medicalization of psychedelics?" I say I worry about the *under*-medicalization of psychedelics. Because it's only medicalization that will do it.

"Half the things I prescribe to people on a daily basis are way more toxic than MDMA," he said in a separate interview. "Yet they're all drugs that are in widespread use."

We are in favor of the slow, boring path as well. This path means more research to better understand the benefits, risks, and trade-offs associated with these and other drugs, and to discover the safest and most effective ways of using them. Yet still today, the potential for such drugs to help relationships is being explored mostly informally, illegally, and through an accumulation of folk knowledge based on anecdotes. Let's bring this work into the light and subject it to the same scientific standards as any other therapeutic intervention involving the use of a chemical substance. There is little to be gained remaining in the dark.

ACKNOWLEDGMENTS

We would like to acknowledge the support of the Uehiro Foundation on Ethics and Education. Without its support, there would be no Chair of Practical Ethics at the University of Oxford, nor the Centre for Practical Ethics. We are deeply grateful for the Foundation's support since 2002, which has literally made this book possible through the creation of positions at the University of Oxford, as well as supporting the Centre and its activities.

Julian is also grateful to the Murdoch Children's Research Institute for providing him with a Visiting Professorial Fellowship in Biomedical Ethics. A substantial part of his work relating to this book was done during a four-month period spent there between July and October 2017, and in the second half of 2018. Julian also benefited from a concurrent Distinguished International Visiting Professorship at the Melbourne Law School, University of Melbourne.

Brian has been affiliated with both the Uehiro Centre (as a Research Fellow), and the Hastings Center (first as a Presidential Visiting Scholar and then as Associate Director of the Yale-Hastings Program in Ethics and Health Policy), throughout much of the writing of this book. He is very grateful to Julian for welcoming him into the Uehiro Centre community and for being such a steadfast mentor over the years. And he wishes to thank Millie Solomon and Dan Callahan, president and cofounder of the Hastings Center, respectively, for inviting him to stay for a year-long residency between 2015 and 2016 to read, write, and think about love drugs and other topics. Erik Parens, a Senior Research Scholar at the Hastings Center, was an especially important influence on Brian during his stay, through weekly lunch meetings and many spirited conversations.

Together, we have many people to thank. First, we are grateful to our coauthors on the various "love drugs" papers for contributing their time and expertise to the development of some of these ideas. Deserving of special mention is Anders Sandberg, who wrote the very first paper on this topic with Julian back in 2008, and who has been a crucial coauthor and general guru since Brian took over the lead in 2010. Olga Wudarczyk also contributed immensely, co-authoring three articles with us, including one as first coauthor with Brian. Finally, thanks to Bennett Foddy, Adam Guastella, and Andrew Vierra for joining us on one paper each, with Bennett lending his expertise on the science and philosophy of addiction, Adam keeping us on track as one of the foremost experts on the use of oxytocin in humans, and Andrew providing nuanced insights into the debate on conversion therapy and the rights of persons with minority sexual orientations.

Then a huge thank-you to Abby Rabinowitz and Rhys Southan, who helped with ideas and edits on a number of chapters, and who injected a lot of energy into this book with their fresh thinking as things moved into the final stretch. Speaking of the final stretch, we are grateful to Stuart, Diane, and Morgan Firestein for letting Brian take over their cottage for periods of quiet concentration, and for general encouragement, and to Ole Martin Moen for a summer's worth of intellectual stimulation and company in Oslo and for reading the entire manuscript and providing many helpful comments. Thanks also to our editor at Stanford University Press, Emily-Jane Cohen, as well as our agents at Janklow & Nesbitt—Emma Parry in New York and Rebecca Carter in London—for believing in this project in the first place and helping us through it with such incredible patience. On the UK side, Jonathan de Peyer has been a wonderful champion of the version of the book published by Manchester University Press. We are grateful to Jamie Ackerman, Sabrina Acosta, Giulia Capicotto, Jamie Mestanza, Rebekah Powers, Ryan Riccioni, Lauren Sardi, Ashley Trueman, Rocci Wilkinson, Miriam Wood, and Jeff Wyneken for editorial assistance. And we thank Joe Alessi, Mario Attie-Picker, John Danaher, Candice Delmas, Robin Dembroff, Daniel Do, Margaret Vivienne Fang, Katarzyna Grunt-Mejer, Carrie Jenkins, Kelsi Lindus, Moya Mapps, Geoffrey Miller, Joshua Teperowski Monrad, Yuri Munir, Sven Nyholm, Ben Sessa, Matthew Strother, and Lori Watson for thoughtful feedback on various parts of this manuscript.

Finally, a sincere thank-you to our colleagues, close friends, and families for all your support, understanding, and cheering on as we worked on this manuscript over so many nights and weekends. We could not have done this without you.

♥ ♥ ♥

Various portions of this book have been adapted from the following essays, and any reused material has been included with the permission of the relevant publisher. Throughout the Notes we have tried to flag the specific sentences and paragraphs that were previously published or only lightly edited, but we may have missed some along the way. Here is a comprehensive list of earlier essays from which we have borrowed or adapted some of our own words: "Neuroenhancement of Love and Marriage: The Chemicals between Us," *Neuroethics* 1, no. 1 (2008); "Natural Selection, Childrearing, and the Ethics of Marriage and Divorce: Building a Case for the Neuroenhancement of Human Relationships," *Philosophy and Technology* 25, no. 4 (2012); "If I Could Just Stop Loving You: Anti-love Biotechnology and the Ethics of a Chemical Breakup," *American Journal of Bioethics* 13, no. 11 (2013); "Could Intranasal Oxytocin Be Used to Enhance Relationships? Research Imperatives,

Clinical Policy, and Ethical Considerations," *Current Opinion in Psychiatry* 26, no. 5 (2013); "Brave New Love: The Threat of High-Tech Conversion Therapy and the Bio-oppression of Sexual Minorities," *American Journal of Bioethics: Neuroscience* 5, no. 1 (2014); "Neuroreductionism about Sex and Love," *Think: A Journal of the Royal Institute of Philosophy* 13, no. 38 (2014); "The Medicalization of Love" and "The Medicalization of Love: Response to Critics," *Cambridge Quarterly of Healthcare Ethics* 24, no. 3, (2015), and 25, no. 4 (2016); "Can You Be Gay by Choice?" in *Philosophers Take On the World*, ed. David Edmonds (Oxford University Press, 2016); "Addicted to Love: What Is Love Addiction and When Should It Be Treated?" and "Love Addiction: Reply to Jenkins and Levy," *Philosophy, Psychiatry, and Psychology* 24, no. 1 (2017); "Love Drugs: Why Scientists Should Study the Effects of Pharmaceuticals on Human Romantic Relationships," *Technology in Society* 52, no. 1 (2018); "Sexual Orientation Minority Rights and High-Tech Conversion Therapy," in *Handbook of Philosophy and Public Policy*, ed. David Boonin (Palgrave Macmillan, 2018); "Psychedelic Moral Enhancement," in *Moral Enhancement: Critical Perspectives*, ed. Michael Hauskeller and Lewis Coyne (Cambridge University Press, 2018); "Love and Enhancement Technology," in *The Oxford Handbook of Philosophy of Love*, ed. Christopher Grau and Aaron Smuts (Oxford University Press, 2019).

NOTES

Chapter 1: Revolution

"Oxford ethicists promote": The headline is actually "Oxford ethicist promotes MDMA to combat divorce," but this makes it sound like only one of us was making the relevant arguments. Scotto, "Oxford ethicist promotes MDMA to combat divorce," *Dose Nation* (blog), February 12, 2013, www.dosenation.com/listing.php?id=8715. **The blogger was referring to an interview:** Brian D. Earp, Anders Sandberg, Julian Savulescu, and Ross Andersen, "The Case for Using Drugs to Enhance Our Relationships (and Our Break-Ups)," *The Atlantic*, January 31, 2013, www.theatlantic.com/technology/archive/2013/01/the-case-for-using-drugs-to-enhance-our-relationships-and-our-break-ups/272615/. **The truth is, we were not promoting the use of MDMA:** See Brian D. Earp, Julian Savulescu, and Anders Sandberg, "Should You Take Ecstasy to Improve Your Marriage? Not So Fast," *Practical Ethics* (blog), June 14, 2012, http://blog.practicalethics.ox.ac.uk/2012/06/should-you-take-ecstasy-to-improve-your-marriage-not-so-fast/. **are moving quickly into the center of mainstream medicine:** Stephen Bright and Martin Williams, "Should Australian Psychology Consider Enhancing Psychotherapeutic Interventions with Psychedelic Drugs? A Call for Research," *Australian Psychologist* 53, no. 6 (December 2018): 467–76. **In the *New York Times* alone:** Khaliya, "The Promise of Ecstasy for PTSD" (editorial), *New York Times*, November 3, 2017, www.nytimes.com/2017/11/03/opinion/ecstasy-ptsd.html; Jan Hoffman, "A Dose of a Hallucinogen from a 'Magic Mushroom,' and Then Lasting Peace," *New York Times*, December 1, 2016, www.nytimes.com/2016/ 12/01/health/hallucinogenic-mushrooms-psilocybin-cancer-anxiety-depression.html; Aaron E. Carroll, "Can Psychedelics Be Therapy? Allow Research to Find Out," *The Upshot* (blog), *New York Times*, July 17, 2017, www.nytimes.com/2017/07/17/upshot/can-psychedelics-be-therapy-allow-research-to-find-out.html. **But now that phase 3 clinical trials:** Kai Kupferschmidt, "All Clear for the Decisive Trial of MDMA in PTSD Patients," *Science News*, August 26, 2017, www.sciencemag.org/news/2017/08/all-clear-decisive-trial-ecstasy-ptsd-patients; Kevin Kunzmann, "FDA Approves Landmark Psilocybin Trial for Treatment-Resistant Depression" (News), *MD Magazine*, August 24, 2018, www.mdmag.com/medical-news/fda-approves-landmark-psilocybin-trial-for-treatmentresistant-depression; MAPS, "MAPS Officially Begins Phase 3 Trials of MDMA-Assisted Psychotherapy for PTSD," Multidisciplinary Association for Psychedelic Studies, November 29, 2018, https://maps.org/news/update/7523–newsletter-november-2018. **Phrases like "paradigm shift":** Eduardo Ekman Schenberg, "Psychedelic-Assisted Psychotherapy: A Paradigm Shift in Psychiatric Research and Development," *Frontiers in Pharmacology* 9, no. 733 (2018): 1–11; David E. Nichols, Matthew W. Johnson, and Charles D. Nichols, "Psychedelics as Medicines: An Emerging New Paradigm," *Clinical Pharmacology and Therapeutics* 101, no. 2 (2017): 209–19. **This was two years after the first pilot study:** Michael C. Mithoefer, Mark T. Wagner, Ann T. Mithoefer, Lisa Jerome, and Rick Doblin, "The Safety and Efficacy of ±3, 4–methylenedioxymethamphetamine-Assisted Psychotherapy in Subjects with Chronic, Treatment-Resistant Posttraumatic Stress Disorder: The First Randomized Controlled Pilot Study," *Journal of Psychopharmacology* 25, no. 4 (2011): 439–52. **PTSD is an often disabling condition:** Ronald C. Kessler et al., "Trauma and PTSD in the WHO World Mental Health Surveys," *European Journal of Psychotraumatology* 8, sup. 5 (2017): 1353383. **As the *Washington Post* reports:** William Wan, "Ecstasy Could Be 'Breakthrough' Therapy for Soldiers, Others Suffering from PTSD," *Washington Post*, August 26, 2017, www.washingtonpost.com/national/health-science/ecstasy-could

-be-breakthrough-therapy-for-soldiers-others-suffering-from-ptsd/2017/08/26/009314ca-842f-11e7-b359-15a3617c767b_story.html?noredirect=on&utm_term=.60754ee7d87e. **"ravaged lives and broken up marriages":** Wan, "Ecstasy . . . " **It affects friends, family:** Danny Horesh and Adam D. Brown, "Post-Traumatic Stress in the Family" (editorial), *Frontiers in Psychology* 9, no. 40 (2018): 1–3. **Yet with PTSD, unprocessed traumas:** Katie Collie, Amy Backos, Cathy Malchiodi, and David Spiegel, "Art Therapy for Combat-Related PTSD: Recommendations for Research and Practice," *Art Therapy* 23, no. 4 (2006): 157–64. **"That's why there's so much frustration and interest":** Wan, "Ecstasy . . . " **Current research suggests:** P. Ø. Johansen and T. S. Krebs, "How Could MDMA (Ecstasy) Help Anxiety Disorders? A Neurobiological Rationale," *Journal of Psychopharmacology* 23, no. 4 (2009): 389–91; Robin L. Carhart-Harris et al., "The Effect of Acutely Administered MDMA on Subjective and BOLD-fMRI Responses to Favourite and Worst Autobiographical Memories," *International Journal of Neuropsychopharmacology* 17, no. 4 (2014): 527–40. **Objectively, it causes the release:** Allison A. Feduccia and Michael C. Mithoefer, "MDMA-Assisted Psychotherapy for PTSD: Are Memory Reconsolidation and Fear Extinction Underlying Mechanisms?" *Progress in Neuro-psychopharmacology and Biological Psychiatry* 84, part A (2018): 221–28. **Subjectively, as one writer:** Wan, "Ecstasy . . . " **Findings from the early MDMA studies:** Kupferschmidt, "All Clear . . . " **"Nothing worked for me":** Dave Philipps, "FDA Agrees to New Trials for Ecstasy as Relief for PTSD Patients," *New York Times*, November 29, 2016, www.nytimes.com/2016/11/29/us/ptsd-mdma-ecstasy.html. **After three sessions of therapy:** MAPS, "MDMA-Assisted Psychotherapy Study Participant C.J. Hardin Hosts Q&A on Reddit," Multidisciplinary Association for Psychedelic Studies, July 25, 2014, https://maps.org/news/media/5252–mdma-assisted-psychotherapy-study-participant-cj-hardin-hosts-q-a-on-reddit. **"We think it works as a catalyst":** Philipps, "FDA Agrees . . . " **"The focus," Dr. Mithoefer says:** Natasha Preskey, "Could MDMA Save Your Relationship?" *Elle*, July 21, 2017, www.elleuk.com/life-and-culture/culture/longform/a36937/could-mdma-save-your-relationship/. **MDMA was granted "breakthrough" status:** Janet Burns, "FDA Designates MDMA as 'Breakthrough Therapy' for Post-Traumatic Stress," *Forbes*, August 28, 2017, www.forbes.com/sites/janetwburns/2017/08/28/fda-designates-mdma-as-breakthrough-therapy-for-post-traumatic-stress/. **The following year, a phase 2 clinical trial:** Michael C. Mithoefer, Ann T. Mithoefer, Allison A. Feduccia, Lisa Jerome, Mark Wagner, Joy Wymer, Julie Holland, Scott Hamilton, Berra Yazar-Klosinski, Amy Emerson, and Rick Doblin, "3, 4–methylenedioxymethamphetamine (MDMA)-Assisted Psychotherapy for Post-Traumatic Stress Disorder in Military Veterans, Firefighters, and Police Officers: A Randomised, Double-Blind, Dose-Response, Phase 2 Clinical Trial," *The Lancet Psychiatry* 5, no. 6 (2018): 486–97. See also Dave Philipps, "Ecstasy as a Remedy for PTSD? You Probably Have Some Questions," *New York Times*, May 1, 2018, www.nytimes.com/2018/05/01/us/ecstasy-molly-ptsd-mdma.html. **As for safety, the accumulated evidence suggests:** Allison A. Feduccia, Julie Holland, and Michael C. Mithoefer, "Progress and Promise for the MDMA Drug Development Program," *Psychopharmacology* 235, no. 2 (2018): 561–71. **And the risk of serious harm:** David J. Heal, Jane Gosden, and Sharon L. Smith, "Evaluating the Abuse Potential of Psychedelic Drugs for Medical Use in Humans," *Neuropharmacology* 142 (2018): 89–115. See also Linda R. Gowing, Susan M. Henry-Edwards, Rodney J. Irvine, and Robert L. Ali, "The Health Effects of Ecstasy: A Literature Review," *Drug and Alcohol Review* 21, no. 1 (2002): 53–63. **Often the legal or medical status:** David Nutt, "Mind-altering Drugs and Research: From Presumptive Prejudice to a Neuroscientific Enlightenment?" *EMBO Reports* 15, no. 3 (2014): 208–11. See also Katherine R. Bonson, "Regulation of Human Research with LSD in the United States (1949–1987)," *Psychopharmacology* 235, no. 2 (2018): 591–604. **journalist Michael Pollan's recent meditation:** Michael Pollan, *How to Change Your Mind: What the New Science of Psychedelics Teaches Us about Consciousness, Dying, Addiction, Depression, and Transcendence* (New York: Penguin, 2014). **The results, published in the *Journal of Psychopharmacology*:** Roland R. Griffiths, Matthew W. Johnson, William A. Richards, Brian D. Richards, Robert Jesse, Katherine A. MacLean, Frederick S. Barrett, Mary P. Cosimano, and Maggie A. Klinedinst, "Psilocybin-Occasioned Mystical-Type Experience in Combination with Meditation and Other Spiritual Practices Produces Enduring Positive Changes in Psychological Functioning and in Trait Measures of Prosocial Attitudes and Behaviors," *Journal* *of Psychopharmacology* 32, no. 1 (2018): 49–69. **Some drugs that are used for medicine:** Andrew Kolodny, David T. Courtwright, Catherine S. Hwang, Peter Kreiner, John L. Eadie, Thomas W. Clark, and G. Caleb Alexander, "The Prescription Opioid and Heroin Crisis: A Public Health Approach to an Epidemic of Addiction," *Annual Review of Public Health* 36 (2015): 559–74. **But other drugs:** Matthew W. Johnson, Roland R. Griffiths, Peter S. Hendricks, and Jack E. Henningfield, "The Abuse Potential of Medical Psilocybin According to the 8 Factors of the Controlled Substances Act," *Neuropharmacology* 142 (2018): 143–66. **That's another headline:** Alex Williams, "How LSD Saved One Woman's Marriage," *New York Times*, January 7, 2017, www.nytimes.com/2017/01/07/style/microdosing-lsd-ayelet-waldman-michael-chabon-marriage.html. **The reference is to a self-experiment:** Ayelet Waldman, *A Really Good Day: How Microdosing Made a Mega Difference in My Mood, My Marriage, and My Life* (New York: Knopf, 2017). **The effects are supposed to be subperceptual:** Erin Brodwin, "Scientists Are About to Find Out How Silicon Valley's LSD Habit Really Affects Productivity," *Business Insider*, June 28, 2017, www.businessinsider.com/microdosing-lsd-effects-risks-science-2017-6. **"I was suffering":** Waldman, *Really Good Day*, xxii. **Her frustration:** Waldman, *Really Good Day*, 31. **Not only can microdosing land you in jail:** The limited data that do exist come from studies that are not very well controlled. See Petter Grahl Johnstad, "Powerful Substances in Tiny Amounts: An Interview Study of Psychedelic Microdosing," *Nordic Studies on Alcohol and Drugs* 35, no. 1 (2018): 39–51; Luisa Prochazkova, Dominique P. Lippelt, Lorenza S. Colzato, Martin Kuchar, Zsuzsika Sjoerds, and Bernhard Hommel, "Exploring the Effect of Microdosing Psychedelics on Creativity in an Open-Label Natural Setting," *Psychopharmacology* 235, no. 12 (2018): 3401–13. See also Thomas Anderson and Rotem Petranker, "'Microdosers' of LSD and Magic Mushrooms Are Wiser and More Creative," *The Conversation*, November 5, 2018, https://theconversation.com/microdosers-of-lsd-and-magic-mushrooms-are-wiser-and-more-creative-101302. **One of us (Brian):** This is a completely fictionalized account of an exchange that did take place. **Some would argue that Sofia:** C. S. I. Jenkins, "'Addicted'? to 'Love'?" *Philosophy, Psychiatry, and Psychology* 24, no. 1 (2017): 93–96. **This is a normative definition of love:** For a discussion of this argument, see Brian D. Earp, Bennett Foddy, Olga A. Wudarczyk, and Julian Savulescu, "Love Addiction: Reply to Jenkins and Levy," *Philosophy, Psychiatry, and Psychology* 24, no. 1 (2017): 101–3; see also Brian D. Earp and Julian Savulescu, "Is There Such a Thing as a Love Drug? Reply to McGee," *Philosophy, Psychiatry, and Psychology* 23, no. 2 (2016): 93–96. **If the feelings between individuals:** bell hooks, *All About Love* (New York: Harper, 2000). **Once we start defining for other people:** See Kristina Gupta, "Why Not a Mannequin? Questioning the Need to Draw Boundaries around Love When Considering the Ethics of 'Love-Altering' Technologies," *Philosophy, Psychiatry, and Psychology* 23, no. 2 (2016): 97–100. For a response, see Andrew McGee, "Love's Exemplars: A Response to Gupta, Earp, and Savulescu," *Philosophy, Psychiatry, and Psychology* 23, no. 2 (2016): 101–2. **Then listen to the incredulous groaning:** See "Christopher Hitchens: Homosexuality Is a Form of Love," www.youtube.com/watch?v=bq4mzH9wqjY. **The tendency to "medicalize" love:** See Earp and Savulescu, "Is There Such a Thing . . . " **One implication of this account:** Berit Brogaard, *On Romantic Love: Simple Truths about a Complex Emotion* (Oxford: Oxford University Press, 2014). **Another prominent theory:** This theory is due to Carrie Jenkins. See Carrie Jenkins, *What Love Is: And What It Could Be* (New York: Basic Books, 2017). **It acknowledges that our capacity for love:** Daphne Blunt Bugental, "Acquisition of the Algorithms of Social Life: A Domain-Based Approach," *Psychological Bulletin* 126, no. 2 (2000): 187–219; Gurit E. Birnbaum and Harry T. Reis, "Evolved to Be Connected: The Dynamics of Attachment and Sex over the Course of Romantic Relationships," *Current Opinion in Psychology* 25, no. 1 (2019): 11–15. **One way Sofia could try:** See Brogaard, *On Romantic Love* . . . **Or maybe they see parallels with Aldous Huxley's:** Aldous Huxley, *Brave New World* (London: Vintage, 1932/1998). **As an online commenter wrote:** See the comments section in Tracy Moore, "Would You Take a Pill to Stay Happily Married?" *Jezebel*, June 12, 2013, https://jezebel.com/would-you-take-a-pill-to-stay-happily-married-512366792. **Whenever society is faced with:** For extended discussions of this perspective as it relates to other technologies, see John Danaher, Brian D. Earp, and Anders Sandberg, "Should We Campaign against Sex Robots?" in *Robot Sex: Social and Ethical Implications*, ed. John Danaher and Neil McArthur

(Cambridge, MA: MIT Press, 2017), 47–71; John Danaher, Sven Nyholm, and Brian D. Earp, "The Quantified Relationship," *American Journal of Bioethics* 18, no. 2 (2018): 3–19; John Danaher, Sven Nyholm, and Brian D. Earp, "The Benefits and Risks of Quantified Relationship Technologies: Response to Open Peer Commentaries on 'The Quantified Relationship,'" *American Journal of Bioethics* 8, no. 2 (2018): W3–W6. **This is a blind spot in Western medicine:** Our argument was spelled out originally in Brian D. Earp and Julian Savulescu, "Love Drugs: Why Scientists Should Study the Effects of Pharmaceuticals on Human Romantic Relationships," *Technology in Society* 52 (2018): 10–16. Larry Young has made a similar argument: Larry J. Young, "Love: Neuroscience Reveals All," *Nature* 457, no. 7226 (2009): 148. **Controlled studies are already underway:** See Sarah Boesveld, "Chemical Seduction: How 'Love Drugs' May One Day Help Couples Save Failing Relationships," *National Post*, March 30, 2013, http://news.nationalpost.com/news/chemical-seduction-how-love-drugs-may-one-day-help-couples-save-failing-relationships; also Adam Guastella, pers. comm., April 13, 2017. **Others may think that in the future:** Gavin G. Enck and Jeanna Ford, "A Responsibility to Chemically Help Patients with Relationships and Love?" *Cambridge Quarterly of Healthcare Ethics* 24, no. 4 (2015): 493–96.

Chapter 2: Love's Dimensions

Among the Yusufzai Pukhtun: See Charles Lindholm, "Leatherworkers and Love Potions," *American Ethnologist* 8, no. 3 (1981): 512–25. Please note that this paragraph is adapted from Julian Savulescu and Anders Sandberg, "Neuroenhancement of Love and Marriage: The Chemicals between Us," *Neuroethics* 1, no. 1 (2008): 31–44. **In Swedish folklore:** Savulescu and Sandberg, "Neuroenhancement of Love and Marriage," 31. **So would drinking plenty of water:** This sentence is adapted from Brian D. Earp, Olga A. Wudarczyk, Anders Sandberg, and Julian Savulescu, "If I Could Just Stop Loving You: Anti-Love Biotechnology and the Ethics of a Chemical Breakup," *American Journal of Bioethics* 13, no. 11 (2013): 3–17. The original source for these claims is Lawrence Babb, "The Physiological Conception of Love in the Elizabethan and Early Stuart Drama," *Publications of the Modern Language Association of America* 56, no. 4 (1941): 1020–35. **Such potions, he counseled:**

Ovid, *Remedia amoris*. An English translation by A. S. Kline with a cool visual layout by Nikolas Schiller is available at www.nikolasschiller.com/blog/index.php/archives/2008/04/03/1345/. **Some scientists think that the more we understand:** See, e.g., Larry J. Young, "Being Human: Love: Neuroscience Reveals All," *Nature* 457, no. 7226 (2009): 148. This sentence is adapted from Brian D. Earp, "Love and Enhancement Technology," in *The Oxford Handbook of Philosophy of Love*, ed. Christopher Grau and Aaron Smuts (Oxford: Oxford University Press, 2019). Please note that much of the material in this chapter is adapted, with permission, from this essay. **In practice drugs are usually thought of:** For examples, see the "drug" entry at https://en.oxforddictionaries.com/definition/drug or https://dictionary.cambridge.org/us/dictionary/english/drug. **A popular textbook on pharmacology:** Jerrold S. Meyer and Linda F. Quenzer, *Psychopharmacology: Drugs, the Brain, and Behavior* (Sunderland, MA: Sinauer Associates, 2004), 4. **In this respect we agree with:** Carrie Jenkins, *What Love Is: And What It Could Be* (New York: Basic Books, 2017). **Although you may have heard that:** Lubomir Lamy, "The Day Love Was Invented," *Psychology Today*, April 17, 2011, www.psychologytoday.com/us/blog/the-heart-it-all/201104/the-day-love-was-invented. **In no special order:** These statements are paraphrased from Sven Nyholm and Lily E. Frank, "From Sex Robots to Love Robots: Is Mutual Love with a Robot Possible?" in *Robot Sex: Social and Ethical Implications,* ed. John Danaher and Neil McArthur (Cambridge, MA: MIT Press, 2017), 219–43. We thank Sven Nyholm for recommending that we highlight these notions early on. **First, though, we'll give a quick overview**: This section is adapted from Earp, "Love and Enhancement Technology . . . " **In rough outline, the science goes like this:** This paragraph and the next are adapted from Brian D. Earp, Anders Sandberg, and Julian Savulescu, "The Medicalization of Love," *Cambridge Quarterly of Healthcare Ethics* 24, no. 3 (2015): 323–36. **These have been described differently:** Helen E. Fisher, Arthur Aron, Debra Mashek, Haifang Li, and Lucy L. Brown, "Defining the Brain Systems of Lust, Romantic Attraction, and Attachment," *Archives of Sexual Behavior* 31, no. 5 (2002): 413–19. **At a more personal level:** For a fascinating recent discussion of one such possible "alternative lifestyle," see Sophie Hemery, "Can Relationship Anarchy Create a World without Suffering?" *Aeon*, November 13, 2018, https://aeon.co/ideas/can-relationship-anarchy-create-a-world-without-heartbreak. **On the contrary, research across many:** Lisa M. Diamond, "What Does Sexual Orientation Orient? A Biobehavioral Model Distinguishing Romantic Love and Sexual Desire," *Psychological Review* 110, no. 1 (2003): 173–92, 174; internal references omitted. **At the same time, she continues, ignoring:** Diamond, "What Does Sexual Orientation Orient?" **Consider the *Mona Lisa*:** The following material is adapted from the appendix to Brian D. Earp and Julian Savulescu, "Love Drugs: Why Scientists Should Study the Effects of Pharmaceuticals on Human Romantic Relationships," *Technology in Society* 52 (2018): 10–16. **What's on the screen:** Jenkins, *What Love Is*, 84 (in advance copy). **Similarly, when it comes to love:** Jenkins, *What Love Is*, 84 (in advance copy). **There are also chastity belts, genital mutilations:** Note that not all genital mutilations or modifications are primarily motivated by a desire to control sexuality; contrary to common belief, they serve different functions and follow different motivations in different societies. For in-depth discussions, see Brian D. Earp, "Between Moral Relativism and Moral Hypocrisy: Reframing the Debate on 'FGM,'" *Kennedy Institute of Ethics Journal* 26, no. 2 (2016): 105–44, E1–E28; Brian D. Earp, "Female Genital Mutilation and Male Circumcision: Toward an Autonomy-Based Ethical Framework," *Medicolegal and Bioethics* 5, no. 1 (2015): 89–104; and Brian D. Earp and Rebecca Steinfeld, "Gender and Genital Cutting: A New Paradigm," in *Gifted Women, Fragile Men*, ed. T. G. Barbat (Euromind Monographs—2) (Brussels: ALDE Group-EU Parliament, 2017). **principle of Default Natural Ethics:** This paragraph and other minor portions of this chapter are adapted from Brian D. Earp, Anders Sandberg, and Julian Savulescu, "Natural Selection, Childrearing, and the Ethics of Marriage (and Divorce): Building a Case for the Neuroenhancement of Human Relationships," *Philosophy and Technology* 25, no. 4 (2012): 561–87. **1947 treatise *Man for Himself*:** Erich Fromm, *Man for Himself: An Enquiry into the Psychology of Ethics* (London: Routledge, 1949), 31. **Our biological nature places certain limitations:** Daphne Blunt Bugental, "Acquisition of the Algorithms of Social life:

A Domain-Based Approach," *Psychological Bulletin* 126, no. 2 (2000): 187–219. Focusing on childrearing and socialization, Bugental writes:

> Deliberate efforts to manage the experiences, behaviors, and values of the young operate within a wide range of flexibility. However, the ease with which such management can occur depends on the extent to which such efforts are consistent or inconsistent with the basic [evolved] algorithms of social life. Many seemingly rational strategies for regulating social relationships may meet a wall of resistance when they are in some way inconsistent with domain algorithms. Socialization or community programs that recognize the basic algorithms of social life (and the optional ways in which those algorithms may be implemented) are more likely to be successful than those that are based on "rational," domain-insensitive principles. (209)

Sexual repression is a key example: Christopher Ryan, "Sexual Repression: The Malady That Considers Itself the Remedy," *Psychology Today*, April 20, 2010, www.psychologytoday.com/blog/sex-dawn/201004/sexual-repression-the-malady-considers-itself-the-remedy. **It has been argued that puritanical:** Kathrin F. Stanger-Hall and David W. Hall, "Abstinence-Only Education and Teen Pregnancy Rates: Why We Need Comprehensive Sex Education in the U.S.," *PLOS ONE* 6, no. 10 (2011): e24658; John S. Santelli et al., "Abstinence-Only-Until-Marriage: An Updated Review of U.S. Policies and Programs and Their Impact," *Journal of Adolescent Health* 61, no. 3 (2017): 273–80. See also Brian D. Earp, "People Are Terrified of Sex," *The Atlantic*, November 12, 2015, www.theatlantic.com/health/archive/2015/11/the-stigma-of-sex-related-health-risks/415518/. **Even the moral crisis of child sexual abuse:** Jason Berry, *Lead Us Not into Temptation: Catholic Priests and the Sexual Abuse of Children* (New York: Doubleday, 1992); Thomas G. Plante, "Catholic Priests Who Sexually Abuse Minors: Why Do We Hear So Much Yet Know So Little?" *Pastoral Psychology* 44 (1996): 305–10; Christopher Ryan and Cacilda Jethá, *Sex at Dawn* (New York: HarperCollins, 2010); Kathleen McPhillips, "The Royal Commission Investigates Child Sexual Abuse: Uncovering Cultures of Sexual Violence in the Catholic Church," in *Rape Culture, Gender Violence, and Religion*. Edited by Caroline Blyth, Emily Colgan, and Katie B. Edwards. (Basingstoke, UK: Palgrave Macmillan, 2018), 53–71; BBC, "Australia Child Abuse Inquiry Finds 'Serious Failings,'" BBC, December 15, 2017, https://www.bbc.com/news/world-australia-42361874; Riazat Butt, "Archbishop Links

Priestly Celibacy and Catholic Sex Abuse Scandals," *The Guardian*, March 11, 2010, https://www.theguardian.com/world/2010/mar/11/priestly-celibacy-catholic-sex-scandals. For a contrary perspective disputing the proposed causative link between celibacy and sexual abuse of minors—from a report commissioned by the United States Conference of Catholic Bishops—see John Jay College of Criminal Justice, "The Nature and Scope of Sexual Abuse of Minors by Catholic Priests and Deacons in the United States, 1950–2002," February 2004, http://www.usccb.org/issues-and-action/child-and-youth-protection/upload/the-nature-and-scope-of-sexual-abuse-of-minors-by-catholic-priests-and-deacons-in-the-united-states-1950-2002.pdf. **Some researchers have described such behavior:** Richard Wrangham and Dale Peterson, *Demonic Males: Apes and the Origins of Human Violence* (London: Houghton Mifflin Harcourt, 1997); Randy Thornhill and Craig T. Palmer, *A Natural History of Rape: Biological Bases of Sexual Coercion* (Cambridge, MA: MIT Press, 2001). **Obviously, this perspective is controversial:** Jerry A. Coyne and Andrew Berry, "Rape as an Adaptation: Is This Contentious Hypothesis Advocacy, Not Science?" *Nature* 404, no. 6774 (2000): 121–22; Jonathan Marks, "Demonic Males: Apes and the Origins of Human Violence," *Human Biology* 71, no. 1 (1999): 143–46. **(Among other problems:** For a general discussion, see Kate Manne, *Down Girl: The Logic of Misogyny* (Oxford, UK: Oxford University Press, 2017). **when these are central to understanding the more immediate causes of rape:** Hilary Rose, "Debating Rape," *The Lancet* 357, no. 9257 (2001): 727–28; Elisabeth A. Lloyd, "Science Gone Astray: Evolution and Rape," *Michigan Law Review* 99, no. 6 (2001): 1536–59. **It is wrong because it is a gross violation:** David Archard, "The Wrong of Rape," *Philosophical Quarterly* 57, no. 228 (2007): 374–93. For further discussion, see Brian D. Earp, "'Legitimate Rape,' Moral Coherence, and Degrees of Sexual Harm," *Think* 14, no. 41 (2015): 9–20; Brian D. Earp, "The Child's Right to Bodily Integrity," in *Ethics and the Contemporary World*, ed. David Edmonds (New York: Routledge, 2019); and the Brussels Collaboration on Bodily Integrity, "Medically Unnecessary Genital Cutting and the Rights of the Child: Moving Toward Consensus," *American Journal of Bioethics*, in press. **So, for example, the finding that males with pedophilia:** R. Blanchard, J. M. Cantor, and L. K. Robichaud, "Biological Factors in the

Development of Sexual Deviance and Aggression in Males," in *The Juvenile Sex Offender* (New York: Guilford Press, 2006), 77–104. **Let us even assume that pedophilia:** Michael C. Seto, "Is Pedophilia a Sexual Orientation?" *Archives of Sexual Behavior* 41, no. 1 (2012): 231–36. **Moreover, if a person with pedophilia volunteered:** For discussions, see Ole Martin Moen, "The Ethics of Pedophilia," *Etikk i Praksis* 9, no. 1 (2015): 111–24; Brian D. Earp, "Pedophilia and Child Sexual Abuse Are Two Different Things and Confusing Them Is Harmful to Children," *Journal of Medical Ethics Blog*, November 11, 2017, http://blogs.bmj.com/medical-ethics/2017/11/11/pedophilia-and-child-sexual-abuse-are-two-different-things-confusing-them-is-harmful-to-children/; Paul J. Fedoroff, "Can People with Pedophilia Change? Yes They Can!" *Current Sexual Health Reports* 10, no. 4 (2018): 207–12; and James M. Cantor, "Can Pedophiles Change?" *Current Sexual Health Reports* 10, no. 4 (2018): 203–6. **The "naturalistic fallacy":** Julia Tanner, "The Naturalistic Fallacy," *Richmond Journal of Philosophy* 13 (2006): 1–6. See also Brian D. Earp, "Science Cannot Determine Human Values," *Think* 15, no. 43 (2016): 17–23. **But something's being natural or biological:** For a nice discussion, with many useful examples, see Steven Pinker, *The Blank Slate: The Modern Denial of Human Nature* (New York: Penguin, 2003). **It is true that nature has great beauty:** See Nick Bostrom and Anders Sandberg, "The Wisdom of Nature: An Evolutionary Heuristic for Human Enhancement," in *Human Enhancement*, ed. Julian Savulescu and Nick Bostrom (Oxford: Oxford University Press, 2009), 375–416. **But nature also allows for great ugliness:** Ole Martin Moen, "The Ethics of Wild Animal Suffering," *Etikk i Praksis* 10, no. 1 (2016): 91–104. **The basic blueprint for our bodies:** Richard Dawkins, *The Blind Watchmaker: Why the Evidence of Evolution Reveals a Universe without Design* (New York: Norton, 1996). **Much of this improvement is due:** Steven Pinker, *Enlightenment Now: The Case for Reason, Science, Humanism, and Progress* (New York: Penguin, 2018). For criticism, see, e.g., Alison Gopnik, "When Truth and Reason Are No Longer Enough," *The Atlantic*, April 2018, www.theatlantic.com/magazine/archive/2018/04/steven-pinker-enlightenment-now/554054/. See also John Gray, "Unenlightened Thinking," *New Statesman*, February 22, 2018, www.newstatesman.com/culture/books/2018/02/unenlightened-thinking-steven-pinker-s-embarrassing-new

-book-feeble-sermon. **In short, without knowing the specifics of a situation:** C. A. J. Coady, "Playing God," in *Human Enhancement*, ed. Julian Savulescu and Nick Bostrom (Oxford: Oxford University Press, 2009), 155–80. **Consider the situation of a person:** For an excellent overview of the philosophy of trans identities and related matters, see Talia Mae Bettcher, "Feminist Perspectives on Trans Issues," in *The Stanford Encyclopedia of Philosophy*, ed. Edward N. Zalta, Spring 2014, https://plato.stanford.edu/entries/feminism-trans/. **whose stable and deeply rooted gender identity:** See Robin Dembroff, "Beyond Binary: Genderqueer as a Critical Gender Kind", *Philosophers' Imprint*, in press. On the issue of misalignment, however, see Robin Dembroff, "Moving Beyond Mismatch," *American Journal of Bioethics* 19, no. 2 (2019): 60–63. **(They have often been subjected to violence:** Rebecca L. Stotzer, "Violence against Transgender People: A Review of United States Data," *Aggression and Violent Behavior* 14, no. 3 (2009): 170–79; Mark A. Walters, Jennifer Paterson, Rupert Brown, and Liz McDonnell, "Hate Crimes against Trans People: Assessing Emotions, Behaviors, and Attitudes toward Criminal Justice Agencies," *Journal of Interpersonal Violence* (2017), online ahead of print at http://journals.sagepub.com/doi/abs/10.1177/0886260517715026; Talia Mae Bettcher, "Evil Deceivers and Make-Believers: On Transphobic Violence and the Politics of Illusion," *Hypatia* 22, no. 3 (2007): 43–65. **Imagine that you are a woman:** Lori Watson, "The Woman Question," *Transgender Studies Quarterly* 3, nos. 1–2 (2016): 246–53. **She writes:** Watson, "Woman Question," 247. **All this because your body is socially:** Watson, "Woman Question," 248. **In order to simply survive in the world:** Sari L. Reisner, Asa Radix, and Madeline B. Deutsch, "Integrated and Gender-Affirming Transgender Clinical Care and Research," *Journal of Acquired Immune Deficiency Syndromes* 72, Suppl. 3 (2016): S235. **hormone treatments or surgeries is the best option for some transgender people:** Jason M. Weissler, Brian L. Chang, Martin J. Carney, David Rengifo, Charles A. Messa, David B. Sarwer, and Ivona Percec, "Gender-Affirming Surgery in Persons with Gender Dysphoria," *Plastic and Reconstructive Surgery* 141, no. 3 (2018): 388e-96e; Elizabeth A. Dietz, "Gender, Identity, and Bioethics," *Hastings Center Report* 46, no. 4 (2016): inside front cover; Stephen B. Levine, "Ethical Concerns about Emerging Treatment Paradigms for

Gender Dysphoria," *Journal of Sex and Marital Therapy* 44, no. 1 (2018): 29–44; Alex Dubov and Liana Fraenkel, "Facial Feminization Surgery: The Ethics of Gatekeeping in Transgender Health," *American Journal of Bioethics* 18, no. 12 (2018): 3–9. **And they may legitimately pursue these means:** Larry R. Martinez and Michelle R. Hebl, "Additional Agents of Change in Promoting Lesbian, Gay, Bisexual, and Transgendered Inclusiveness in Organizations," *Industrial and Organizational Psychology* 3, no. 1 (2010): 82–85. **Certainly, no one should be forced:** At least in most cases, no one should be forced; however, the moral bioenhancement of violent psychopaths has been proposed as a possible exception to this general rule. See Elvio Baccarini and Luca Malatesti, "The Moral Bioenhancement of Psychopaths," *Journal of Medical Ethics* 43, no. 10 (2017): 697–701. **To summarize, how well our lives go:** See, e.g., Julian Savulescu, "Genetic Interventions and the Ethics of Enhancement of Human Beings," in *The Oxford Handbook of Bioethics*, ed. Bonnie Steinbock (Oxford: Oxford University Press, 2009), 417–30; and Julian Savulescu, "New Breeds of Humans: The Moral Obligation to Enhance," *Reproductive BioMedicine Online* 10 (2005): 36–39.

Chapter 3: Human Natures

Consider a married couple: Portions of this chapter are adapted, with permission, from Brian D. Earp, "Love and Enhancement Technology," in *The Oxford Handbook of Philosophy of Love*, ed. Christopher Grau and Aaron Smuts (Oxford: Oxford University Press, 2019). **A plausible answer is, it depends:** Christopher Ryan and Cacilda Jethá, *Sex at Dawn: The Prehistoric Origins of Modern Sexuality* (New York: HarperCollins, 2010); Dossie Easton and Janet W. Hardy, *The Ethical Slut: A Practical Guide to Polyamory, Open Relationships, and Other Adventures* (Berkeley: Celestial Arts, 2010); Terri D. Conley, Ali Ziegler, Amy C. Moors, Jes L. Matsick, and Brandon Valentine, "A Critical Examination of Popular Assumptions about the Benefits and Outcomes of Monogamous Relationships," *Personality and Social Psychology Review* 17, no. 2 (2013): 124–41; but see George W. Dent Jr., "Traditional Marriage: Still Worth Defending," *Brigham Young University Journal of Public Law* 18, no. 2 (2003): 419–47. **Some philosophers argue:** Bryan R. Weaver and Fiona Woollard, "Marriage and the Norm of Monogamy,"

The Monist 91, nos. 3/4 (2008): 506–22. **Instead, for most couples, and for society:** The following quotes from Dan Savage are from a profile by Mark Oppenheimer, "Married, with Infidelities," *New York Times*, June 30, 2011, www.nytimes.com/2011/07/03/magazine/infidelity-will-keep-us-together.html. **As one religious scholar:** John Witte Jr., "Why Monogamy Is Natural," *On Faith*, October 2, 2012, www.onfaith.com/onfaith/2012/10/02/why-monogamy-is-natural/12105. **According to the prominent evolutionary theorists:** David P. Barash and Judith Eve Lipton, *The Myth of Monogamy: Fidelity and Infidelity in Animals and People* (New York: Macmillan, 2002), 8. **Yet even social monogamy turns out to be:** Agustin Fuentes, "Re-evaluating Primate Monogamy," *American Anthropologist* 100, no. 4 (1998): 890–907. **"Like bonobos and chimps":** Ryan and Jethá, *Sex at Dawn*, 2. **But not everyone is on board:** According to Geoffrey Miller, only a very small percentage of mainstream evolutionary psychology, anthropology, or biology researchers agree with Ryan and Jethá that our ancestors were full-fledged bonobo-like polyamorists; but he grants that many—perhaps the majority—of researchers working on human mating have what he calls "monogamist biases." Geoffrey Miller, pers. comm., November 3, 2018; all quotes from Miller with permission. **"blue is the natural eye color for humans:** Carrie Jenkins, *What Love Is: And What It Could Be* (New York: Basic Books 2017), 93 (in advance copy). **Just as some people may be "wired up":** See Brian D. Earp, "Can You Be Gay by Choice?" in *Philosophers Take on the World*, ed. David Edmonds (Oxford: Oxford University Press, 2016), 95–98. **then polyamory is probably natural for some people, too:** For a fascinating discussion, see Ann Tweedy, "Polyamory as a Sexual Orientation," *University of Cincinnati Law Review* 79, no. 4 (2011): 1461–1515. **"Every mating system:** Geoffrey Miller, pers. comm., November 3, 2018. **As empirical studies are beginning to suggest:** Terri D. Conley, Jes L. Matsick, Amy C. Moors, and Ali Ziegler, "Investigation of Consensually Non-monogamous Relationships: Theories, Methods, and New Directions," *Perspectives on Psychological Science* 12, no. 2 (2017): 205–32. **Chances are—in this day and age:** For an excellent discussion of some of the stigmas facing polyamorous individuals, especially black men, see Justin Leonard Clardy, "I Don't Want to Be a Player No More: An Exploration of the Denigrating Effects of 'Player' as a

Stereotype against African American Polyamorous Men," *AnAlize Journal of Gender and Feminist Studies* 11, no. 25 (2018): 38–59. For related ideas see Eric Anderson, *The Monogamy Gap: Men, Love, and the Reality of Cheating* (Oxford: Oxford University Press, 2011). **Jealousy has deep evolutionary roots:** David M. Buss, *The Evolution of Desire* (New York: Basic Books, 2016). **Could a biological jealousy inhibitor:** This section and some latter parts of this chapter are adapted from Earp, "Love and Enhancement Technology . . . " **"A 54-year-old male manager:** Gordon Parker and Elaine Barrett, "Morbid Jealousy as a Variant of Obsessive-Compulsive Disorder," *Australian and New Zealand Journal of Psychiatry* 31, no. 1 (1997): 133–38. **"Marital difficulties followed his:** Parker and Barrett, "Morbid Jealousy . . . ," 134. **"Personality review suggested a man:** Parker and Barrett, "Morbid Jealousy . . . ," 134. **"an almost universal behavior among the jealous:** Paul E. Mullen, "A Phenomenology of Jealousy," *Australian and New Zealand Journal of Psychiatry* 24 (1990): 17–28. **Four weeks later, the man reported:** Parker and Barrett, "Morbid Jealousy . . . ," 134. **In fact, this man wasn't diagnosed:** Parker and Barrett, "Morbid Jealousy . . . ," 133. **Could science one day help you:** For an extremely thoughtful set of arguments to the effect that biochemically enhancing fidelity, while most likely morally permissible in many cases, may nevertheless be imprudent for some couples (because it could make it harder to tell whether one's partner is disposed not to sleep with others for the right kinds of reasons, that is, reasons that matter for genuinely attenuating one's vulnerability in a romantic relationship), see Robbie Arrell, "Should We Biochemically Enhance Sexual Fidelity?" *Royal Institute of Philosophy Supplements* 83 (2018): 389–414. **Unlike many antidepressant medications:** Jack G. Modell, Charles R. Katholi, Judith D. Modell, and R. Lawrence DePalma, "Comparative Sexual Side Effects of Bupropion, Fluoxetine, Paroxetine, and Sertraline," *Clinical Pharmacology and Therapeutics* 61, no. 4 (1997): 476–87. **According to one study, people treated with SSRI-based:** Modell et al., "Comparative Sexual Side Effects . . . ," 476. **Evidence for this prediction comes from:** Steven D. Levitt, "Heads or Tails: The Impact of a Coin Toss on Major Life Decisions and Subsequent Happiness," *National Bureau of Economic Research Working Paper 22487*, August 2016, www.nber.org/papers/w22487. **Once we have the power to alter a situation:** See

Gerald Dworkin, "Is More Choice Better Than Less?" *Midwest Studies in Philosophy* 7 (1982): 47–61. **On the biological side, we looked:** Miller pointed out to us that there are also "many lifestyle approaches that people unconsciously use to manage their libidos. After marriage, many guys put on abdominal fat ('dad bod'), which reduces testosterone levels and libido. Many women also put on weight, which tends to reduce libido; if they feel a mismatch to their husband's libido, they might take up exercise, which tends to increase libido." Geoffrey Miller, pers. comm., November 3, 2018.

Chapter 4: Little Heart-Shaped Pills

"madly dote": William Shakespeare, "A Midsummer Night's Dream" (*Project Gutenberg*, 1595/1986), http://nfs.sparknotes.com/msnd/page_42.html. **Consider the 1960s chart-topper:** "Love Potion No. 9," as recorded by The Searchers. Lyrics from https://genius.com/The-scarchers-love-potion-no-9–lyrics. **"*Amortentia*:** J. K. Rowling, *Harry Potter and the Half-Blood Prince* (London: Bloomsbury, 2005). Quote from "Love Potions: Hogwarts' Most Intoxicating Tonic," www.pottermore.com/features/love-potion-guide-hogwarts-most-intoxicating-tonic. **For one thing, love drugs:** See our discussion in Brian D. Earp, Ander Samberg, and Julian Savulescu, "The Medicalization of Love: Response to Critics," *Cambridge Quarterly of Healthcare Ethics* 25, no. 4 (2016): 759–71. **"From our earliest years":** Kayt Sukel, *Dirty Minds* (New York: Free Press, 2012), 56. **"Hormones actually control sexual behavior:** Sukel, *Dirty Minds*, 56. **"Hormones are not absolute regulators:** Quoted in Sukel, *Dirty Minds*, 57. **"As you grow up:** Helen Fisher, *Anatomy of Love: A Natural History of Mating, Marriage, and Why We Stray* (New York: Norton, 2016), 318–19. Partially against this view, as one (anonymous) reader of our manuscript has speculated, at least some drugs, namely psychedelics, "*can* actually induce mental model transformations; they can change what matters to you, in life and in relationships. Perhaps induced plasticity in combination with therapeutic suggestion could change not just the quality of your feelings, but the way you direct your feelings." For some suggestive evidence, see Calvin Ly et al., "Psychedelics Promote Structural and Functional Neural Plasticity," *Cell Reports* 23, no. 11 (2018): 3170–82. **"Better relationships through chemistry":** For the original slogan and context, see https://en.wikipedia.org/wiki/Better_Living_Through_Chemistry. **Something that requires choice:** In this we agree with Erich Fromm in *The Art of Loving* (New York: Harper, 1956). Please note that this sentence and several of the following are adapted from Brian D. Earp, "Love and Enhancement Technology," in *The Oxford Handbook of Philosophy of Love,* ed. Christopher Grau and Aaron Smuts (Oxford: Oxford University Press, 2017). **the idea that love "takes work":** This point was raised in a wonderful essay by Erik Parens, "On Good and Bad Forms of Medicalization," *Bioethics* 27, no. 1 (2013): 28–35. **These neurochemicals, in fact:** Brian D. Earp, Olga A. Wudarczyk, Bennett Foddy, and Julian Savulescu, "Addicted to Love: What Is Love Addiction and When Should It Be Treated?" *Philosophy, Psychiatry, and Psychology* 24, no. 1 (2017): 77–92; Brian D. Earp, Olga A. Wudarczyk, Bennett Foddy, and Julian Savulescu, "Love Addiction: Reply to Jenkins and Levy," *Philosophy, Psychiatry, and Psychology* 24, no. 1 (2017): 101–3. **Instead, erotic touch, sex, and orgasm:** For a general discussion, see Cindy Hazan and Lisa M. Diamond, "The Place of Attachment in Human Mating," *Review of General Psychology* 4, no. 2 (2000): 186–204. **You can even order oxytocin off the internet:** We can't vouch for any of the stuff you can get off the internet. It might not even be real oxytocin, or if it is, there's no guarantee it has been prepared in such a way that it would have any effect. **These factors have been shown to play:** John M. Gottman, "Psychology and the Study of Marital Processes," *Annual Review of Psychology* 49, no. 1 (1998): 169–97; Thomas Ledermann, Guy Bodenmann, Myriam Rudaz, and Thomas N. Bradbury, "Stress, Communication, and Marital Quality in Couples," *Family Relations* 59, no. 2 (2010): 195–206. **Another issue is that some people really struggle:** For a related argument, see John Danaher, Sven Nyholm, and Brian D. Earp, "The Quantified Relationship," *American Journal of Bioethics* 18, no. 2 (2018): 3–19. **"I just rounded the corner:** From the comments section of Tracy Moore, "Would You Take a Pill to Stay Happily Married?" *Jezebel*, June 12, 2013, http://jezebel.com/would-you-take-a-pill-to-stay-happily-married-512366792. **"The hard thing about constant relationships:** From the comments section of Moore, "Would You Take a Pill . . . " **And now Addyi (flibanserin):** For a discussion of why the "female Viagra" moniker is misleading, see Alice

G. Walton, "Why Libido Drug Addyi Is Not the 'Female Viagra,'" *Forbes*, August 19, 2015, www.forbes.com/sites/alicegwalton/2015/08/19/fda-approves-addyi-but-it-is-not-the-female-viagra/#4e22a00d2889. **is being controversially touted as a prolibido:** On the controversy, see Weronika Chańska and Katarzyna Grunt-Mejer, "The Unethical Use of Ethical Rhetoric: The Case of Flibanserin and Pharmacologisation of Female Sexual Desire," *Journal of Medical Ethics* 42, no. 11 (2016): 701–4; Antonie Meixel, Elena Yanchar, and Adriane Fugh-Berman, "Hypoactive Sexual Desire Disorder: Inventing a Disease to Sell Low Libido," *Journal of Medical Ethics* 41, no. 10 (2015): 859–62. **Testosterone blockers, for example:** Thomas Douglas, Pieter Bonte, Farah Focquaert, Katrien Devolder, and Sigrid Sterckx, "Coercion, Incarceration, and Chemical Castration: An Argument from Autonomy," *Journal of Bioethical Inquiry* 10, no. 3 (213): 393–405. **For example, some drugs used to treat depression:** Adam Opbroek, Pedro L. Delgado, Cindi Laukes, Cindy McGahuey, Joanna Katsanis, Francisco A. Moreno, and Rachel Manber, "Emotional Blunting Associated with SSRI-Induced Sexual Dysfunction: Do SSRIs Inhibit Emotional Responses?" *International Journal of Neuropsychopharmacology* 5, no. 2 (2002): 147–51; W. Jason Barnhart, Eugene H. Makela, and Melissa J. Latocha, "SSRI-Induced Apathy Syndrome: A Clinical Review," *Journal of Psychiatric Practice* 10, no. 3 (2004): 196–99; Randy A. Sansone and Lori A. Sansone, "SSRI-Induced Indifference," *Psychiatry* 7, no. 10 (2010): 14–18. **To illustrate, we'll pick one main feature:** This portion of the chapter is adapted from Earp, "Love and Enhancement Technology . . . " **Consider the view that true love:** Kevin E. Hegi and Raymond M. Bergner, "What Is Love? An Empirically-Based Essentialist Account," *Journal of Social and Personal Relationships* 27, no. 5 (2010): 620–36. See also Bennett Helm, "Love," in *The Stanford Encyclopedia of Philosophy*, August 11, 2017, https://plato.stanford.edu/archives/fall2017/entries/love/ (especially the section on love as "robust concern"). **But there are plenty of case reports:** Opbroek et al., "Emotional Blunting Associated with SSRI-Induced Sexual Dysfunction . . . ," 147. **Fully 80 percent of the patients:** Opbroek et al., "Emotional Blunting Associated with SSRI-Induced Sexual Dysfunction. . . ," 147. For more recent evidence and discussion, see Elcin Ozsin Aydemir, Eda Aslan, and Mustafa Kemal Yazici, "SSRI Induced Apathy Syndrome,"

Psychiatry and Behavioral Sciences 8, no. 2 (2018): 63–70. **Indeed, on some views, the experience:** See Neil Delaney, "Romantic Love and Loving Commitment: Articulating a Modern Ideal," *American Philosophical Quarterly* 33, no. 4 (1996): 339–56. **In other words, wanting to be physically intimate:** For a more complex account that questions the line between platonic and romantic love, see Laurence Dumortier, "Anarchic Intimacies: Queer Friendships and Erotic Bonds," doctoral dissertation, University of California, Riverside, 2017, https://escholarship.org/uc/item/76510324. **Clearly, those things:** See Hichem Naar, "Real-World Love Drugs: Reply to Nyholm," *Journal of Applied Philosophy* 33, no. 2 (2016): 197–201. **But there have also been controlled:** Barry T. Jones, Ben C. Jones, Andy P. Thomas, and Jessica Piper, "Alcohol Consumption Increases Attractiveness Ratings of Opposite-Sex Faces: A Possible Third Route to Risky Sex," *Addiction* 98, no. 8 (2003): 1069–75; Nick Neave, Carmen Tsang, and Nick Heather, "Effects of Alcohol and Alcohol Expectancy on Perceptions of Opposite-Sex Facial Attractiveness in University Students," *Addiction Research and Theory* 16, no. 4 (2008): 359–68. **For a twist on the usual finding:** Laurent Bègue, Brad J. Bushman, Oulmann Zerhouni, Baptiste Subra, and Medhi Ourabah, "'Beauty Is in the Eye of the Beer Holder': People Who Think They Are Drunk Also Think They Are Attractive," *British Journal of Psychology* 104, no. 2 (2013): 225–34. **This book deals with nudges and probabilities:** For a critical discussion of this "nudges and probabilities" claim, see Michael Hauskeller, "Clipping the Angel's Wings: Why the Medicalization of Love May Still Be Worrying," *Cambridge Quarterly of Healthcare Ethics* 24, no. 3 (2015): 361–65. **a prototypical instance of love:** Beverley Fehr and James A. Russell, "The Concept of Love Viewed from a Prototype Perspective," *Journal of Personality and Social Psychology* 60, no. 3 (1991): 425–39. **This same hormone is released through intimate touch:** Cindy Hazan and Lisa M. Diamond, "The Place of Attachment in Human Mating," *Review of General Psychology* 4, no. 2 (2000): 186–204. Please note that the next few paragraphs are adapted from Earp, "Love and Enhancement Technology . . . " **testosterone, which can be boosted:** Susan R. Davis and Jane Tran, "Testosterone Influences Libido and Well-Being in Women," *Trends in Endocrinology and Metabolism* 12, no. 1 (2001): 33–37. **Scientists should study these effects:** This part of the chapter through to the end is adapted from Brian D.

Earp and Julian Savulescu, "Love Drugs: Why Scientists Should Study the Effects of Pharmaceuticals on Human Romantic Relationships," *Technology in Society* 52, no. 1 (2018): 10–16. **A huge amount of data suggests:** For a review, see Olga A. Wudarczyk, Brian D. Earp, Adam Guastella, and Julian Savulescu, "Could Intranasal Oxytocin Be Used to Enhance Relationships? Research Imperatives, Clinical Policy, and Ethical Considerations," *Current Opinion in Psychiatry* 26, no. 5 (2013): 474–84. **Close relationships also show up:** See, e.g., Derek Parfit, *Reasons and Persons* (Oxford: Oxford University Press, 1984); and James Griffin, *Well-Being: Its Meaning, Measurement, and Moral Importance* (Oxford: Clarendon Paperbacks, 1986). **And some philosophers argue:** Sven Nyholm, "The Medicalization of Love and Narrow and Broad Conceptions of Human Well-Being," *Cambridge Quarterly of Healthcare Ethics* 24, no. 3 (2015): 337–46. **Likewise, if other drugs:** Here we are using the term "therapeutic" in its sense of having a positive effect on the body or mind, or contributing to a feeling or state of well-being. This is as opposed to the sense of "healing a disease," whereby a specific pathology must be present. For further discussion, see Brian D. Earp, Anders Sandberg, and Julian Savulescu, "The Medicalization of Love," *Cambridge Quarterly of Healthcare Ethics* 24, no. 3 (2015): 323–36. **Many drug-based treatments:** For a thoughtful discussion, see Nina L. Etkin, "Side Effects: Cultural Constructions and Reinterpretations of Western Pharmaceuticals," *Medical Anthropology Quarterly* 6, no. 2 (1992): 99–113. **These are commonly understood:** See the entry "side effect," www.merriam-webster.com/dictionary/side%20effect. **It has become obvious that:** Neil Levy, Thomas Douglas, Guy Kahane, Sylvia Terbeck, Philip J. Cowen, Miles Hewstone, and Julian Savulescu, "Are You Morally Modified? The Moral Effects of Widely Used Pharmaceuticals," *Philosophy, Psychiatry, and Psychology* 21, no. 2 (2014): 111–25. **This is because the chemical properties:** Helen Fisher and J. Anderson Thomson Jr., "Lust, Romance, Attachment: Do the Side Effects of Serotonin-Enhancing Antidepressants Jeopardize Romantic Love, Marriage, and Fertility?" in *Evolutionary Cognitive Neuroscience*, ed. Steven Platek, Julian Paul Keenan, and Todd Shackelford (Cambridge, MA: MIT Press, 2007), 245. **Between 2005 and 2008:** Laura A. Pratt, Debra J. Brody, and Qiuping Gu, "Antidepressant Use in Persons Aged 12 and Over: United States, 2005–2008," *NCHS Data Brief No. 76*, October 2011, www.sefap.it

/servizi_letteratu racardio_201110/db76.pdf. **with SSRIs being the most commonly prescribed:** Mayo Clinic Staff, "Depression (Major Depressive Disorder)," *Mayo Clinic Online*, July 9, 2013, www.mayoclinic.org/diseases-conditions/depression/basics/definition/con-20032977. **"He soon experienced diminished:** Fisher and Thomson, "Lust, Romance, Attachment . . . ," 267. **increased levels of serotonin caused by SSRIs:** Fisher and Thomson, "Lust, Romance, Attachment . . . ," 257. **"After two bouts of depression:** Jerry Frankel, "Reviving Romance," *New York Times*, May 11, 2004, F4. **Consider this very different profile:** Louisa Kamps, "The Couple Who Medicates Together," *Elle*, April 18, 2012, www.elle.com/life-love/sex-relationships/advice/a14208/the-couple-who-medicates-together-654677/. **All that scientists can say at this point:** Fisher and Thomson, "Lust, Romance, Attachment . . . ," 269. **According to the U.S. Centers for Disease Control and Prevention:** Jo Jones, William Mosher, and Kimberly Daniels, "Current Contraceptive Use in the United States, 2006–2010, and Changes in Patterns of Use since 1995," *National Health Statistics Reports* 60 (2012): 1–25. **Combined methods work by suppressing:** John D. Jacobson "Birth Control Pills—Overview," in *A.D.A.M. Medical Encyclopedia*, U.S. National Library of Medicine, 2018. **As women who use birth control know:** Margaret F. McCann and Linda S. Potter, "Progestin-Only Oral Contraception: A Comprehensive Review: X. Common Side Effects," *Contraception* 50, no. 6 (1994): S114–S138. **Some forms of hormonal birth control:** Mary Wood Littleton, "The Truth about 'The Pill' and Your Sex Drive," *WebMD*, January 28, 2015, www.webmd.com/sex/birth-control/features/the-pill-and-desire. **While the libido-altering effects:** See, e.g., Alexandra Alvergne and Virpi Lummaa, "Does the Contraceptive Pill Alter Mate Choice in Humans?" *Trends in Ecology and Evolution* 25, no. 3 (2010): 171–79. See also Trond Viggo Grøntvedt, Nicholas M. Grebe, Leif Edward Ottesen Kennair, and Steven W. Gangestad, "Estrogenic and Progestogenic Effects of Hormonal Contraceptives in Relation to Sexual Behavior: Insights into Extended Sexuality," *Evolution and Human Behavior* 38, no. 3 (2017): 283–92. For a recent review and discussion, see Santiago Palacios and Mariella Lilue, "Hormonal Contraception and Sexuality," *Current Sexual Health Reports* 10, no. 4 (2018): 345–52. **One common theory goes:** Karl Grammer, Bernhard Fink, and Nick Neave, "Human Pheromones and

Sexual Attraction," *European Journal of Obstetrics and Gynecology and Reproductive Biology* 118, no. 2 (2005): 135–42; however, see Wendy Wood, Laura Kressel, Priyanka D. Joshi, and Brian Louie, "Meta-analysis of Menstrual Cycle Effects on Women's Mate Preferences," *Emotion Review* 6, no. 3 (2014): 229–49, which is largely nonsupportive of this view. **For example, women who are using oral contraception:** S. Craig Roberts and Anthony C. Little, "Good Genes, Complementary Genes and Human Mate Preferences," *Genetica* 132, no. 3 (2008): 309–21. See also S. Craig Roberts, "Effect of Birth Control on Women's Preferences," in *Encyclopedia of Evolutionary Psychological Science*, ed. Todd K. Shackelford and Viviana Weekes-Shackelford (Cham, Switzerland: Springer, 2019); and Andrea C. Gore, Amanda M. Holley, and David Crews, "Mate Choice, Sexual Selection, and Endocrine-Disrupting Chemicals," *Hormones and Behavior* 101 (2018): 3–12. **The psychologist S. Craig Roberts:** S. Craig Roberts, Kateřina Klapilová, Anthony C. Little, Robert P. Burriss, Benedict C. Jones, Lisa M. DeBruine, Marion Petrie, and Jan Havlíček, "Relationship Satisfaction and Outcome in Women Who Meet Their Partner While Using Oral Contraception," *Proceedings of the Royal Society of London B: Biological Sciences* 279, no. 1732 (2012): 1430–36. **Their study found that women who:** Roberts et al., "Relationship Satisfaction and Outcome in Women . . . ," 1430. **Curiously, "the same women:** Roberts et al., "Relationship Satisfaction and Outcome in Women . . . ," 1430. **Either way, the authors concluded:** Roberts et al., "Relationship Satisfaction and Outcome in Women . . . ," 1430. For an up-to-date discussion of findings (and nonfindings) in this area, see Patrick Jern, Antti Kärnä, Janna Hujanen, Tatu Erlin, Annika Gunst, Helmi Rautaheimo, Emilia Öhman, S. Craig Roberts, and Brendan P. Zietsch, "A High-Powered Replication Study Finds No Effect of Starting or Stopping Hormonal Contraceptive Use on Relationship Quality," *Evolution and Human Behavior* 39, no. 4 (2018): 373–79. See also Ruben C. Arslan, Katharina M. Schilling, Tanja M. Gerlach, and Lars Penke, "Using 26,000 Diary Entries to Show Ovulatory Changes in Sexual Desire and Behavior," *Journal of Personality and Social Psychology*, August 27, 2018, online ahead of print, http://psycnet.apa.org/record/2018-41799-001. **In her article:** All quotes from Kamps, "The Couple Who Medicates Together . . . "

Chapter 5: Good-Enough Marriages

In her book *Marriage Confidential*: This paragraph is adapted with permission from Brian D. Earp, "Love and Enhancement Technology," in *The Oxford Handbook of Philosophy of Love*, ed. Christopher Grau and Aaron Smuts (Oxford: Oxford University Press, 2019). **It isn't high-distress:** Pamela Haag, *Marriage Confidential* (New York: HarperCollins, 2011), xi. **Instead, she says, it's relatively low-conflict:** This characterization is from marriage researcher Paul R. Amato and colleagues. See, e.g., Paul Amato and Bryndl Hohmann-Marriott, "A Comparison of High-and Low-Distress Marriages That End in Divorce," *Journal of Marriage and Family* 69, no. 3 (2007): 621–38. **Something like what we described:** This paragraph, and other material following in this chapter, is adapted from Earp, "Love and Enhancement Technology . . . " **If you are a spouse with:** Haag, *Marriage Confidential*, xiii. **"the problem that has no name":** Betty Friedan, *The Feminine Mystique* (New York: Norton, 2010). **Yet there are life values:** This paragraph is adapted from Brian D. Earp. "The Ethics of Infant Male Circumcision," invited lecture, Uehiro Seminar Series from University of Oxford, Oxford, England, June 7, 2013. **Ethical approaches based on community:** Eva Feder Kittay, "The Ethics of Care, Dependence, and Disability," *Ratio Juris* 24, no. 1 (2011): 49–58. **Our ability to be autonomous:** Bruce Jennings, "Reconceptualizing Autonomy: A Relational Turn in Bioethics," *Hastings Center Report* 46, no. 3 (2016): 11–16; Carol Gilligan, "Hearing the Difference: Theorizing Connection," *Hypatia* 10, no. 2 (1995): 120–27; Virginia Held, *Justice and Care: Essential Readings in Feminist Ethics* (New York: Routledge, 2018). **In a nutshell:** Paul R. Amato, "Good Enough Marriages: Parental Discord, Divorce, and Children's Long-Term Well-Being," *Virginia Journal of Social Policy and the Law* 9 (2001): 71–94, 71. **Of course, women are usually:** Sarah Schoppe-Sullivan, "Dads are More Involved in Parenting, Yes, but Moms Still Put in More Work," *The Conversation*, February 2, 2017, https://theconversation.com/dads-are-more-involved-in-parenting-yes-but-moms-still-put-in-more-work-72026; Suman Bhattacharyya, "Women Still Do the Lion's Share of the Housework, but the Gap Is Narrowing," *Fiscal Times*, June 26, 2016, www.thefiscaltimes.com/2016/06/26/Women-Still-Do-Lion-s-Share-Housework-Gap-Narrowing. For a recent discussion of the importance of publicly provided childcare for gender justice, see Vidhi Chhaochharia, "The Tale of Two Germanies Highlights How Childcare Provision Benefits Women," *LSE Business Review* (2018), http://eprints.lse.ac.uk/89466/1/businessreview-2018-04-19-the-tale-of-two-germanies-highlights-how.pdf. See also Mi Young An and Ito Peng, "Diverging Paths? A Comparative Look at Childcare Policies in Japan, South Korea and Taiwan," *Social Policy and Administration* 50, no. 5 (2016): 540–58. **Sometimes they become oblivious:** Judith S. Wallerstein, Julia M. Lewis, and Sandra Blakeslee, *The Unexpected Legacy of Divorce: A Twenty-Five Year Landmark Study* (New York: Hyperion, 2000), xx. Research in this area is of course controversial; see Michelle Jaffee, "Experts Split over Divorce's Legacy for Children," *Chicago Tribune*, August 15, 2004, www.chicagotribune.com/news/ct-xpm-2004-08-15-0408150450-story.html. For a fascinating discussion of some of the deeper philosophy of science (e.g., methodological and epistemological) issues, see Elizabeth Anderson, "Uses of Value Judgements in Science: A General Argument, with Lessons from a Case Study of Feminist Research on Divorce," *Hypatia* 19, no. 1 (2004): 1–24. **Many of them fear:** Wallerstein et al., *Unexpected Legacy of Divorce*, xiii. **"For some children:** Amato, "Good Enough Marriages . . . ," 73–74.

Chapter 6: Ecstasy as Therapy

As she later told a journalist: Mike Sheffield, "Could MDMA Save Your Relationship?" *Complex Magazine*, February 17, 2016, www.complex.com/life/2016/02/how-molly-could-help-your-relationship. All quotes from Autumn are from this source. **In the 1980s, before it:** Ben Sessa, "Is There a Case for MDMA-Assisted Psychotherapy in the UK?" *Journal of Psychopharmacology* 21, no. 2 (2017): 220–24. Please note that the next several paragraphs and some other portions of this chapter are adapted from Brian D. Earp, "Psychedelic Moral Enhancement," *Royal Institute of Philosophy Supplement* 83 (2018): 415–39. **Writing in the *Journal of Psychoactive*:** George R. Greer and Requa Tolbert, "A Method of Conducting Therapeutic Sessions with MDMA," *Journal of Psychoactive Drugs* 30, no. 4 (1998): 371–79, 375. **"We never recommended:** Greer and Tolbert, "Method of Conducting . . . ," 372. **and that of other pioneers:** For a nice history of the pioneering work of Greer,

Tolbert, and many others, see Torsten Passie, "The Early Use of MDMA ('Ecstasy') in Psychotherapy (1977–1985)," *Drug Science, Policy and Law* 4, (2018): 1–19. **Especially at lower doses:** Patrick Vizeli and Matthias E. Liechti, "Safety Pharmacology of Acute MDMA Administration in Healthy Subjects," *Journal of Psychopharmacology* 31, no. 5 (2017): 576–88. **According to Greer and Tolbert:** Greer and Tolbert, "Method of Conducting . . . ," 378. **mescaline, a cactus-derived drug:** Peter N. Jones, "The Native American Church, Peyote, and Health: Expanding Consciousness for Healing Purposes," *Contemporary Justice Review* 10, no. 4 (2007): 411–25; John H. Halpern et al., "Psychological and Cognitive Effects of Long-Term Peyote Use among Native Americans," *Biological Psychiatry* 58, no. 8 (2005): 624–31. **Mescaline and MDMA have:** Harold Kalant, "The Pharmacology and Toxicology of 'Ecstasy' (MDMA) and Related Drugs," *Canadian Medical Association Journal* 165, no. 7 (2001): 917–28. **"The mescaline experience is:** The quote finishes with "to everyone, but especially intellectuals." Aldous Huxley, *The Doors of Perception* (London: Chatto and Windus, 1954). See page 53 of the online version, http://nacr.us/media/text/the_doors_of_perception.pdf. **"One cannot take psilocybin:** William A. Richards, "Understanding the Religious Import of Mystical States of Consciousness Facilitated by Psilocybin," in *The Psychedelic Policy Quagmire: Health, Law, Freedom, and Society*, ed. J. Harold Ellens and Thomas B. Roberts (Denver: Praeger, 2015), 139–44, 140. **More than a century ago:** William James, "Subjective Effects of Nitrous Oxide," *Mind* 7, no. 1 (1882): 186–208. **subjective changes "may determine attitudes:** William James, *The Varieties of Religious Experience* (Mineola, NY: Dover, 1902), 379. **Think of a magnifying glass:** Thanks to Ole Martin Moen for this analogy. **One woman told us:** This is a fictional story based on conversations we have had with actual people. **In 1985 the U.S. government:** National Institute of Drug Abuse, "What Is the History of MDMA?" 2017, www.drugabuse.gov/publications/research-reports/mdma-ecstasy-abuse/what-is-the-history-of-mdma. **On July 27, 1984:** The account here, through "catch-22" is closely paraphrased from Passie, "Early Use of MDMA . . . ," 11–12. **The FDA based its reasoning:** Passie, "Early Use of MDMA . . . ," 14. **And it really is a sorry situation:** Ben Sessa, *The Psychedelic Renaissance:*

Reassessing the Role of Psychedelic Drugs in 21st Century Psychiatry and Society (Herndon, VA: Muswell Hill Press, 2012). See also Evan Wood, Daniel Werb, Brandon D. L. Marshall, Julio S. G. Montaner, and Thomas Kerr, "The War on Drugs: A Devastating Public-Policy Disaster," *The Lancet* 373, no. 9668 (2009): 989–90. **One such story comes from:** Don Lattin, "Marriage Is Driving Some to Drugs and It May Not Be a Bad Thing," *California Magazine*, February 13, 2017, https://alumni.berkeley.edu/california-magazine/just-in/2017-02-14/marriage-driving-some-drugs-and-it-may-not-be-bad-thing. The subsequent quotes from George Greer are also from this source. **Specifically, we risk "falsifying:** Leon R. Kass et al., *Beyond Therapy: Biotechnology and the Pursuit of Happiness* (Washington, DC: President's Council on Bioethics, 2003), 227. **Of course, the way our emotions work:** Thanks to Kelsi Lindus for the examples after "or simply." **"Where I live, they love fireworks:** Quoted in "'Like a Hug from Everyone Who Loves You'—How MDMA Could Help Patients with Trauma," *Pharmaceutical Journal* 301, no. 7918 (October 2018), www.pharmaceutical-journal.com/news-and-analysis/features/like-a-hug-from-everyone-who-loves-you-how-mdma-could-help-patients-with-trauma/20205586.article?firstPass=false. All subsequent quotes from Lubecky are also from this source. Please note that the original punctuation in the quote beginning "I wouldn't talk about the trauma" has been slightly edited to improve readability. **In fact, the journalist Mike Sheffield:** Sheffield, "Could MDMA Save Your Relationship?" **The thought here is that:** As Julian and our colleague Anders Sandberg wrote in the very first paper on relationship neuroenhancement back in 2008:

> In the case of love between two people, there is usually some form of compatibility, some shared values, some event or aspect of personality that enables and leads to the love. The feeling has an "autobiographical anchor," making it authentic. Again, it is important to distinguish between the use of love potions to create new love and to foster existing love. The use of drugs to instill a new love is more likely to create inauthentic love, since the causal reasons for the love may lie in the drug (and external events surrounding the situation), rather than the particular person loved. This would not be the case in an established loving relationship that is losing its momentum. (Julian Savulescu and Anders Sandberg, "Neuroenhancement of Love and Marriage: The Chemicals Between Us," *Neuroethics* 1, no. 1 [2008]: 31–44, 40)

Please note that some of this chapter is also adapted from this essay. **Another concern about authenticity:** Portions of this section are adapted from Brian D. Earp, Anders Sandberg, and Julian Savulescu, "The Medicalization of Love," *Cambridge Quarterly of Healthcare Ethics* 24, no. 3 (2015): 323–36. **We recently did some research:** Brian D. Earp, Joshua A. Skorburg, Jim A. C. Everett, and Julian Savulescu, "Addiction, Identity, Morality," *AJOB: Empirical Bioethics* 10, no. 2 (2019): 136–53. See also Kevin P. Tobia, "Personal Identity and the Phineas Gage Effect," *Analysis* 75, no. 3 (2015): 396–405; Kevin P. Tobia, "Personal Identity, Direction of Change, and Neuroethics," *Neuroethics* 9, no. 1 (2016): 37–43; and Kevin P. Tobia, "Changes Becomes You," *Aeon*, September 19, 2017, https://aeon.co/essays/to-be-true-to-ones-self-means-changing-to-become-that-self. **The philosopher Marya Schechtman:** Marya Schechtman, "Empathic Access: The Missing Ingredient in Personal Identity," *Philosophical Explorations* 4, no. 2 (2001): 95–111. For more work on transformative experiences, see L. A. Paul, *Transformative Experience* (Oxford: Oxford University Press, 2014). We wish to thank Rebecca Bamford for first alerting us to Schechtman's work in this context; Rebecca Bamford, "Unrequited: Neurochemical Enhancement of Love," *Cambridge Quarterly of Healthcare Ethics* 24, no. 3 (2015): 355–60, 359. **Some people have had horrible:** Andrew C. Parrott, Luke A. Downey, Carl A. Roberts, Cathy Montgomery, Raimondo Bruno, and Helen C. Fox, "Recreational 3, 4–Methylenedioxymethanmphetamine or 'Ecstasy': Current Perspective and Future Research Projects," *Journal of Psychopharmacology* 31, no. 8 (2017): 959–66. **Some have died from:** John A. Henry, K. J. Jeffreys, and Shelia Dawling, "Toxicity and Deaths from 3, 4–Methylenedioxymethamphetamine ('Ecstasy')," *The Lancet* 340, no. 8816 (1992): 384–87. **"Recreational Ecstasy is often taken with:** Ben Sessa, pers. comm., September 27, 2017. Quoted with consent. All subsequent quotes from Sessa are from this interview, except for those attributed to his published article. **In a recent article:** Sessa, "Is There a Case for MDMA-Assisted Psychotherapy . . . ," 223. **Even so, he concludes that:** Sessa, "Is There a Case for MDMA-Assisted Psychotherapy . . . ," 223, emphasis in original.

Chapter 7: Evolved Fragility

This fundamental mismatch: The basic argument in this chapter is adapted primarily from Julian Savulescu and Anders Sandberg, "Neuroenhancement of Love and Marriage: The Chemicals Between Us," *Neuroethics* 1, no. 1 (2008): 31–44. For further related discussion, see Martin Daly and Margo I. Wilson, "The Evolutionary Psychology of Marriage and Divorce," in *The Ties That Bind: Perspectives on Marriage and Cohabitation*, ed. Linda J. Waite et al. (New York: De Gruyter, 2000), 91–110. See also Daphne Blunt Bugental, "Acquisition of the Algorithms of Social Life: A Domain-Based Approach," *Psychological Bulletin* 126, no. 2 (2000): 187–219. **Rather, our ancestors evolved:** We can make this more precise. Strictly speaking, our ancestors didn't "evolve" anything; rather, the blind process of natural selection, faced with slowly changing circumstances over an incredible stretch of time, eventually favored the survival to reproduction of ancestral parents that tended to stick together rather than mate-and-run, and who passed that tendency on to offspring as part of their genetic inheritance. We use agential, anthropomorphic language for evolution purely for the sake of expediency. **Until the last one hundred years:** This section (including the next several paragraphs) draws heavily from Dario Maestripieri, "The Seven Year Itch: Theories of Marriage, Di vorce, and Love," *Psychology Today*, February 3, 2012, www.psychology-today.com/us/blog/games-primates-play/201202/the-seven-year-itch-theories-marriage-divorce-and-love. His book covering similar material is *Games Primates Play: An Undercover Investigation of the Evolution and Economics of Human Relationships* (New York: Basic Books, 2012). **Human babies are particularly vulnerable:** For a theoretical overview, see Mart R. Gross, "The Evolution of Parental Care," *Quarterly Review of Biology* 80, no. 1 (2005): 37–45. For more recent discussions and alternative perspectives, see Kristina M. Durante, Paul W. Eastwick, Eli J. Finkel, Steven W. Gangestad, and Jeffry A. Simpson, "Pair-Bonded Relationships and Romantic Alternatives: Toward an Integration of Evolutionary and Relationship Science Perspectives," *Advances in Experimental Social Psychology* 53, no. 1 (2016): 1–74; and James F. O'Connell, Kristen Hawkes, Frank W. Marlowe, and Nicholas G. Blurton Jones, "Paternal Investment and Hunter-Gatherer Divorce Rates," in *Adaptation and Human Behavior*, ed. Napoleon Chagnon (Abingdon, UK: Routledge, 2017), 69–90.

To that end, modern hunter-gatherers: For an overview, see Kristen Hawkes, James O'Connell, and Nicholas Blurton Jones, "Hunter-Gatherer Studies and Human Evolution: A Very Selective Review," *American Journal of Physical Anthropology* 165, no. 4 (2018): 777–800. **It has been speculated that:** Maestripieri, "Seven Year Itch." As Geoffrey Miller pointed out to us, modern !Kung live in fairly marginal environments with poor resources. Typical ancestral interbirth intervals may have been more like 2–3 years in more propitious environments. Miller, pers. comm., November 3, 2018. **In the late 1980s, the anthropologist:** Helen Fisher, *Anatomy of Love: A Natural History of Mating, Marriage, and Why We Stray* (New York: Norton, 2016). **She found that when committed partners:** Maestripieri, "Seven Year Itch." See also Helen Fisher, "Evolution of Human Serial Pairbonding," *American Journal of Physical Anthropology* 78, no. 3 (1989): 331–54. **One interpretation of this finding:** Maestripieri, "Seven Year Itch." **albeit a controversial one:** Daly and Wilson have forcefully criticized Fisher's "four-year itch" theory, finding the evidence in support of it to be thin and unconvincing. See Daly and Wilson, "Evolutionary Psychology of Marriage and Divorce," 103. Fisher has defended her theory in various subsequent articles and books, including Helen Fisher, "Planned Obsolescence? The Four-Year Itch," Edge.org, 2008, www.edge.org/response-detail/11507; and Fisher, *Anatomy of Love*. **Human appetite [is] surprisingly elastic:** Michael Pollan, *The Omnivore's Dilemma* (New York: Penguin, 2012), excerpt from the text available at www.pbs.org/pov/foodinc/excerpt-michael-pollans-the-omnivores-dilemma/. **Consider *Playboy*:** David M. Buss, *The Evolution of Desire* (New York: Basic Books, 2016), 103–4. We should note that *Playboy* has undergone various changes since the death of its founder, Hugh Hefner, in 2017. The editorial practices described here are from the pre-2017 version of the magazine, although similar practices may still be in use. See Jessica Bennett, "Will the Millennials Save Playboy?" *New York Times*, August 2, 2019, https://www.nytimes.com/2019/08/02/business/woke-playboy-millennials.html. **What *Playboy*-purchasing men:** Buss, "Evolution of Desire." **For monogamous relationships:** Andrew Greeley, "Marital Infidelity," *Society* 31, no. 4 (1994): 9–13. For a related discussion, see Jennifer S. Hirsch, Sergio Meneses, Brenda Thompson, Mirka Negroni, Blanca Pelcastre, and Carlos Del

Rio, "The Inevitability of Infidelity: Sexual Reputation, Social Geographies, and Marital HIV Risk in Rural Mexico," *American Journal of Public Health* 97, no. 6 (2007): 986–96. See also Padmanabha Ramanujam, Yugank Goyal, and Sriya Sridhar, "Cultural Institutions in New Technology: Evidence from Internet Infidelity," in *Internet Infidelity*, ed. Sanjeev P. Sahni and Garima Jainpp (Singapore: Springer Nature Singapore, 2018), 45–67. **Divorce has now overtaken death to:** See Savulescu and Sandberg, "Neuroenhancement of Love and Marriage." See also Kirsten Gravningen, Kirstin R. Mitchell, Kaye Wellings, Anne M. Johnson, Rebecca Geary, Kyle G. Jones, Soazig Clifton, et al., "Reported Reasons for Breakdown of Marriage and Cohabitation in Britain: Findings from the Third National Survey of Sexual Attitudes and Lifestyles (Natsal-3)," *PLOS ONE* 12, no. 3 (2017): e0174129. **As somebody once said:** This quote is from page 80 of the 2016 edition of Helen Fisher's *Anatomy of Love*. She attributes it to Oscar Wilde, perhaps because it sounds like something Wilde would have said, but the closest thing we could find is from George Bernard Shaw's 1903 drama, *Man and Superman*, spoken by the character Mendoza: "There are two tragedies in life. One is to lose your heart's desire. The other is to gain it." But the plot thickens! According to *Wikiquote* (which may or may not be a reliable source), this line is apparently derived from a *different* quote by Oscar Wilde, from Act 3 of *Lady Windermere's Fan* (1892), spoken by the character Dumby: "In this world there are only two tragedies. One is not getting what one wants, and the other is getting it." That's still not the same thing as what Fisher attributes to Wilde, but some mash-up of the two is pretty close. As an aside, we wonder how many Oscar Wilde quotes have been misattributed to George Bernard Shaw and vice versa. **Once we realize how implausible:** These points are well expressed by Esther Perel, *Mating in Captivity* (New York: HarperCollins, 2007), xiv. **As the anthropologist Donald Symons:** Donald Symons, "Darwinism and Contemporary Marriage," in *Contemporary Marriage*, ed. Kingsley Davis (New York: Russell Sage, 1985), 133–55. **Divorce rates are notoriously hard to pin down:** They also vary considerably depending on individual difference variables, such as age at marriage and education level. According to Geoffrey Miller, college-educated people who get married after their midtwenties, before they have kids, have divorce rates under 25 percent. Higher IQ also predicts lower divorce rate. Geoffrey Miller, pers. comm., November 3, 2018. See also Paul R. Amato, "Research on Divorce: Continuing Trends and New Developments," *Journal of Marriage and Family* 72 (2010): 650–66. Please note that this paragraph is adapted from Brian D. Earp, Anders Sandberg, and Julian Savulescu, "Natural Selection, Childrearing, and the Ethics of Marriage (and Divorce): Building a Case for the Neuroenhancement of Human Relationships," *Philosophy and Technology* 25, no. 4 (2012) 561–87. **This ratio, at least in recent decades:** Belinda Lusombe, "Are Marriage Statistics Divorced from Reality?" *Time*, May 24, 2010, www.time.com/time/magazine/article/0,9171,1989124,00.html. **Basically, there are about twice:** Other ways at arriving at the 50 percent statistic depend on projections from data collected in the 1970s, 1980s, and 1990s based on the questionable assumption that those earlier trends would continue into the present day. Some experts think that divorce rates have been declining in recent years; others think they are rising. But when you take everything together, as the marriage researcher Paul Amato has recently stated, the 50 percent figure "appears to be reasonably accurate." Quoted in Luscombe, "Are Marriage Statistics Divorced . . . " **Passion, at least, was often:** Perel, *Mating in Captivity*, 8. According to Geoffrey Miller, this view of marriage may be distorted by historical bias. He wrote to us: "Among the wealthy elites, marriage was often for political and resource reasons. Among the majority of folks who never make it into the history books, however—e.g., medieval serfs, townspeople, settlers, factory workers, etc., I suspect that falling in love was a very common reason for marriage." Geoffrey Miller, pers. comm., November 3, 2018. Quoted with consent. **and marriage was more about connecting:** Stephanie Coontz, *Marriage: A History* (New York: Penguin, 2005), 6. **In other words, right up until:** Coontz, *Marriage*, 9. **It was too important an economic:** Coontz, *Marriage*, 9. **Four factors have led to:** Coontz, *Marriage*, 307–8, paraphrased. **If they wanted a long-lasting:** Coontz, *Marriage*, 308. **We seem to have created a paradox:** Paraphrased from Perel, *Mating in Captivity*, 8. **By dismantling them, we:** Perel, *Mating in Captivity*, 8–9. **As Coontz notes, we can no more:** Coontz, *Marriage*, 308. At least, we can no longer do these things, realistically, on a wide scale. Obviously, some people may very well turn to subsistence farming and artisanal economic work, and the more people that do these things, the better it would likely be for our planet. Likewise, within smaller communities or subpopulations, alternative norms for relationships that do not place so much weight on a single partner to fulfill romantic and emotional needs may well be sustainable, as we discussed, for example, with polyamory. **"In today's world it's easy:** Judith S. Wallerstein and Sandra Blakeslee, *The Good Marriage* (Boston: Houghton-Mifflin, 1995), 5.

Chapter 8: Wonder Hormone

For almost a decade: All quotes in this paragraph are from, or within, Ed Yong, "One Molecule for Love, Morality, and Prosperity?" *Slate*, July 17, 2012, https://slate.com/technology/2012/07/oxytocin-is-not-a-love-drug-dont-give-it-to-kids-with-autism.html. This chapter is adapted with permission from Olga A. Wudarczyk, Brian D. Earp, Adam Guastella, and Julian Savulescu, "Could Intranasal Oxytocin Be Used to Enhance Relationships? Research Imperatives, Clinical Policy, and Ethical Considerations," *Current Opinion in Psychiatry* 26, no. (2013): 474–84. **On Amazon you can buy:** See www.amazon.com/OxyLuv-Oxytocin-unnatural-preservatives-fillers/dp/B00MEV8BMM. **Still, a range of products:** See for example www.amazon.com/Attrakt-Oxytocin-Pheromone-Spray-Cologne/dp/B00N2600OK. **Critical studies have manipulated:** Mary M. Cho, A. Courtney DeVries, Jessie R. Williams, and C. Sue Carter, "The Effects of Oxytocin and Vasopressin on Partner Preferences in Male and Female Prairie Voles (*Microtus ochrogaster*)," *Behavioral Neuroscience* 113, no. 5 (1999): 1071–79; Thomas R. Insel and Terrence J. Hulihan, "A Gender-Specific Mechanism for Pair Bonding: Oxytocin and Partner Preference Formation in Monogamous Voles," *Behavioral Neuroscience* 109, no. 4 (1995): 782–89; Jessie R. Williams, Thomas R. Insel, Carol R. Harbaugh, and C. Sue Carter, "Oxytocin Administered Centrally Facilitates Formation of a Partner Preference in Female Prairie Voles (*Microtus ochrogaster*)," *Journal of Neuroendocrinology* 6, no. 3 (1994): 247–50; James T. Winslow, Nick Hastings, C. Sue Carter, Carroll R. Harbaugh, and Thomas R. Insel, "Pair Bonding in the Monogamous Prairie Vole: A Role for Central Vasopressin," *Nature* 365 (1993): 545–48. **In other studies, infusing:** L. J. Young, M. M. Lim, B. Gingrich, and T. R. Insel, "Cellular Mechanisms of Social Attachment," *Hormones and Behavior* 40, no.2 (2001):

133–38. **But several researchers have argued:** Helen E. Fisher, Arthur Aron, and Lucy L. Brown, "Romantic Love: A Mammalian Brain System for Mate Choice," *Philosophical Transactions of the Royal Society B: Biological Sciences* 361, no. 1476 (2006): 2173–86; Zoe R. Donaldson and Larry J. Young, "Oxytocin, Vasopressin, and the Neurogenetics of Sociality," *Science* 322, no. 5903 (2008): 900–904. **Mothers who observe a photograph:** Andreas Bartels and Semir Zeki, "The Neural Correlates of Maternal and Romantic Love," *NeuroImage* 21, no. 3 (2004): 1155–66; F. Loup, Elaine Tribollet, M. Dubois-Dauphin, and Jean Jaques Dreifuss, "Localization of High-Affinity Binding Sites for Oxytocin and Vasopressin in the Human Brain: An Autoradiographic Study," *Brain Research* 555, no. 2 (1991): 220–32. **This effect is amplified:** Lane Strathearn, Peter Fonagy, Janet Amico, and P. Read Montague, "Adult Attachment Predicts Maternal Brain and Oxytocin Response to Infant Cues," *Neuropsychopharmacology* 34, no. 13 (2009): 2655–66. **Similar activation patterns happen in adults:** Fisher, Aron, and Brown, "Romantic Love . . . ," 58–62. **Although none of these findings:** Anne Campbell, "Oxytocin and Human Social Behavior," *Personality and Social Psychology Review* 14, no. 3 (2010): 281–95. **This kind of design:** Ted J. Kaptchuk, "The Double-Blind, Randomized, Placebo-Controlled Trial: Gold Standard or Golden Calf?" *Journal of Clinical Epidemiology* 54, no. 6 (2001): 541–49; see also Brian D. Earp, "Mental Shortcuts," *Hastings Center Report* 46, no. 2 (2016), inside front cover; Brian D. Earp, "The Unbearable Asymmetry of Bullshit," *Health Watch* 101 (Spring 2016): 4–5. **In just the last year or so:** Anthony Lane, Oliver Luminet, Gideon Nave, and Moira Mikolajczak, "Is There a Publication Bias in Behavioural Intranasal Oxytocin Research on Humans? Opening the File Drawer of One Laboratory," *Journal of Neuroendocrinology* 28, no. 4 (2016): 1–15; Simon Oxenham, "Everything You've Heard about Sniffing Oxytocin Might Be Wrong," *New Scientist*, May 16, 2016, www.newscientist.com/article/2088200–everything-youve-heard-about-sniffing-oxytocin-might--be-wrong; Gideon Nave, Colin Camerer, and Michael McCullough, "Does Oxytocin Increase Trust in Humans? A Critical Review of Research," *Perspectives on Psychological Science* 10, no. 6 (2015): 772–89. **Accordingly, "the remarkable reports that:** Hasse Walum, Irwin D. Waldman, and Larry J. Young, "Statistical and Methodological Considerations for the Interpretation of

Intranasal Oxytocin Studies," *Biological Psychiatry* 79, no. 3 (2016): 251–57, 257. **This has become a big issue:** For discussions by Brian and his colleagues, see Brian D. Earp and David Trafimow, "Replication, Falsification, and the Crisis of Confidence in Social Psychology," *Frontiers in Psychology* 6, no. 621 (2015): 1–11; Jim A. C. Everett and Brian D. Earp, "A Tragedy of the (Academic) Commons: Interpreting the Replication Crisis in Psychology as a Social Dilemma for Early-Career Researchers," *Frontiers in Psychology* 6, no. 1152 (2015): 1–4; David Trafimow and Brian D. Earp, "Badly Specified Theories Are Not Responsible for the Replication Crisis in Social Psychology: Comment on Klein," *Theory and Psychology* 26, no. 4 (2016): 540–48; Brian D. Earp, Jim A. C. Everett, Elizabeth N. Madva, and J. Kiley Hamlin, "Out, Damned Spot: Can the 'Macbeth Effect' Be Replicated?" *Basic and Applied Social Psychology* 36, no. 1 (2014): 91–98; Brian D. Earp, "Falsification: How Does It Relate to Reproducibility?" *Key Concepts in Research Methods*, ed. Jean-François Morin et al. (Oxford: Oxford University Press, in press); and Etienne LeBel, Randy J. McCarthy, Brian D. Earp, Malte Elson, and Wolf Vanpaemel, "A Unified Framework to Quantify the Credibility of Scientific Findings," *Advances in Methods and Practices in Psychological Science* 1, no. 3, 389–402. **And it isn't just OT research:** John P. Ioannidis, "Why Most Published Research Findings Are False," *PLOS Medicine* 2, no. 8 (2005): e124; Open Science Collaboration, "Estimating the Reproducibility of Psychological Science," *Science* 349, no. 6251 (2015): aac471; Brian D. Earp, "What Did the OSC Replication Initiative Reveal about the Crisis in Psychology?" *BMC Psychology* 4, no. 28 (2016): 1–19; Brian A. Nosek and Timothy M. Errington, "Making Sense of Replications," *Elife* 6, (2017): e23383. See also Florian Cova, Brent Strickland, Angela Abatista, Aurélien Allard, James Andow, Mario Attie, James Beebe, et al., "Estimating the Reproducibility of Experimental Philosophy," *Review of Philosophy and Psychology* (2018): 1–36 (online publication). **The most famous study:** Beate Ditzen, Marcel Schaer, Barbara Gabriel, Guy Bodenmann, Ulrike Ehlert, and Markus Heinrichs, "Intranasal Oxytocin Increases Positive Communication and Reduces Cortisol Levels during Couple Conflict," *Biological Psychiatry* 65, no. 9 (2009): 728–31. **The headline finding:** In terms of the replication issue, at least one recent study has shown similar effects, namely, a reduction in cortisol (among women) following a dyadic conflict task: Julianne

C. Flanagan, Melanie S. Fischer, Paul J. Nietert, Sudie E. Back, Megan Moran-Santa Maria, Alexandra Snead, and Kathleen T. Brady, "Effects of Oxytocin on Cortisol Reactivity and Conflict Resolution Behaviors among Couples with Substance Misuse," *Psychiatry Research* 260 (2018): 346–52. Of course, we don't know how big the "file drawer" is for studies of this kind (that is, the number of studies with a similar design that failed to show an effect and were therefore never submitted for publication), so the need for healthy skepticism remains. **As Ditzen noted about her research:** Quoted in Jayne Dawkins, "Oxytocin: Love Potion #1?" *Elsevier Press Release*, April 29, 2009, www.elsevier.com/about/press-releases/research-and-journals/oxytocin-love-potion-1. The subsequent quotes in this paragraph are also from this source. **Specifically, it has been shown:** Markus Heinrichs, Thomas Baumgartner, Clemens Kirschbaum, and Ulrike Ehlert, "Social Support and Oxytocin Interact to Suppress Cortisol and Subjective Responses to Psychosocial Stress," *Biological Psychiatry* 54, no. 12 (2003): 1389–98; Peter Kirsch, Christine Esslinger, Qiang Chen, Daniela Mier, Stefanie Lis, Sarina Siddhanti, Harald Gruppe, Venkata S. Mattay, Bernd Gallhofer, and Andreas Meyer-Lindenberg, "Oxytocin Modulates Neural Circuitry for Social Cognition and Fear in Humans," *Journal of Neuroscience* 25, no. 49 (2005): 11489–493; Gregor Domes, Markus Heinrichs, Jan Philipp Gläscher, Christian Büchel, Dieter F. Braus, and Sabine C. Herpertz, "Oxytocin Attenuates Amygdala Responses to Emotional Faces Regardless of Valence," *Biological Psychiatry* 62, no. 10 (2007): 1187–90. **increase trust, eye contact:** Adam J. Guastella, Philip Bowden Mitchell, and Mark R. Dadds, "Oxytocin Increases Gaze to the Eye Region of Human Faces," *Biological Psychiatry* 63, no. 1 (2008): 3–5. **mind reading:** Gregor Domes, Markus Heinrichs, Andre Michel, Christoph Berger, and Sabine C. Herpertz, "Oxytocin Improves 'Mind-Reading' in Humans," *Biological Psychiatry* 61, no. 6 (2007): 731–33. **and empathy:** René Hurlemann et al., "Oxytocin Enhances Amygdala-Dependent, Socially Reinforced Learning and Emotional Empathy in Humans," *Journal of Neuroscience* 30, no. 14 (2010): 4999–5007. **easier to remember the good parts:** Markus Heinrichs, Gunther Meinlschmidt, Werner Wippich, Ulrike Ehlert, and Dirk H. Hellhammer, "Selective Amnesic Effects of Oxytocin on Human Memory," *Physiology and Behavior* 83, no. 1 (2004): 31–38; Adam J. Guastella, Phillip B. Mitchell, and Frosso Mathews,

"Oxytocin Enhances the Encoding of Positive Social Memories in Humans," *Biological Psychiatry* 64, no. 3 (2008): 256–58. **In a recent study, OT enhanced:** Ann-Kathrin Kreuder, Lea Wassermann, Michael Wollseifer, Beate Ditzen, Monika Eckstein, Birgit Stoffel-Wagner, Jürgen Hennig, René Hurlemann, and Dirk Scheele, "Oxytocin Enhances the Pain-Relieving Effects of Social Support in Romantic Couples," *Human Brain Mapping* (2018), online ahead of print, https://on linelibrary.wiley.com/doi/full/10.1002/hbm.24368. **In a relationship therapy context:** John C. Norcross, *Psychotherapy Relationships That Work* (Oxford: Oxford University Press, 2011); Sandra Murray, John G. Holmes, Dale Wesley Griffin, "The Benefits of Positive Illusions: Idealization and the Construction of Satisfaction in Close Relationships," *Journal of Personality and Social Psychology* 70, no. 1 (1996): 79–98. **They also had less of a reflexive:** Dirk Scheele, Nadine Striepens, Onur Güntürkün, Sandra Deutschlander, Wolfgang Maier, Keith M. Kendrick, and Rene Hurlemann, "Oxytocin Modulates Social Distance between Males and Females," *Journal of Neuroscience* 32, no. 46 (2012): 16074–79. **In line with this view:** Christian Unkelbach, Adam J. Guastella, and Joseph P. Forgas, "Oxytocin Selectively Facilitates Recognition of Positive Sex and Relationship Words," *Psychological Science* 19, no. 11 (2008): 1092–94. **More generally, OT may make:** Adam J. Guastella, Philip B. Mitchell, and Frosso Mathews, "Oxytocin Enhances the Encoding of Positive Social Memories in Humans," *Biological Psychiatry* 64, no. 3 (2008): 256–258. **As the neuroscientist Nadine Striepens:** Nadine Striepens, Keith M. Kendrick, Wolfgang Maier, and Rene Hurlemann, "Prosocial Effects of Oxytocin and Clinical Evidence for Its Therapeutic Potential," *Frontiers in Neuroendocrinology* 32, no. 4 (2011): 426–50, 445. **OT seems to have a dark side:** Ed Yong, "The Dark Side of Oxytocin, Much More Than Just a 'Love Hormone,'" *National Geographic*, November 29, 2010, http://phenomena.nationalgeographic.com/2010/11/29/the-dark-side-of-oxytocin-much-more-than-just-a-love-hormone/. **OT can increase feelings of envy:** Simone Shamay-Tsoory, Meytal Fischer-Shofty, Jonathan Dvash, Hagai Harari, Nufar Perach-Bloom, and Yechiel Levkovitz, "Intranasal Administration of Oxytocin Increases Envy and Schadenfreude (Gloating)," *Biological Psychiatry* 66, no. 9 (2009): 864–70. **It also seems that people behave more:** Carolyn H. Declerck, Christophe Boone, and Toko Kiyonari, "Oxytocin and

Cooperation under Conditions of Uncertainty: The Modulating Role of Incentives and Social Information," *Hormones and Behavior* 57, no. 3 (2010): 368–74. **And there is even evidence that:** Carsten K. W. De Dreu, Lindred L. Greer, Gerben A. Van Kleef, Shaul Shalvi, and Michel J. J. Handraaf, "Oxytocin Promotes Human Ethnocentrism," *Proceedings of the National Academy of Sciences* 108, no. 4 (2011): 1262–66. **These results suggest that OT could:** Greg Miller, "The Promise and Perils of Oxytocin," *Science* 339, no. 6117 (2013): 267–69, 269. **People with borderline personality disorder:** Jennifer Bartz, Daphine Simeon, Holly Hamilton, Suah Kim, Sarah Crystal, Ashley Braun, Victor Vicens, and Eric Hollander, "Oxytocin Can Hinder Trust and Cooperation in Borderline Personality Disorder," *Social Cognitive and Affective Neuroscience* 6, no. 5 (2011): 556–63. **And people with an anxious or insecure attachment:** Jennifer A. Bartz, Jamil Zaki, Kevin N. Ochsner, Niall Bolger, Alexander Kolevzon, Natasha Ludwig, and John Lydon, "Effects of Oxytocin on Recollections of Maternal Care and Closeness," *Proceedings of the National Academy of Sciences* 107, no. 50 (2010): 21371–75. **Finally, OT seems to improve empathic:** Jennifer A. Bartz, Jamil Zaki, Niall Bolger, Eric Hollander, Natasha N. Ludwig, Alexander Kolevzon, and Kevin N. Ochsner, "Oxytocin Selectively Improves Empathic Accuracy," *Psychological Science* 21, no. 10 (2010): 1426–28. **How it influences thoughts:** Miller, "Promise and Perils. . . ," 269. See also Adam J. Guastella and Colin Macleod, "A Critical Review of the Influence of Oxytocin Nasal Spray on Social Cognition in Humans: Evidence and Future Directions," *Hormones and Behavior* 61, no. 3 (2012): 410–18, 416. For a more recent study looking at a moderating influence of intimate partner violence, see Amber Jarnecke, Eileen Barden, Sudie E. Back, Kathleen Brady, and Julianne C. Flanagan, "Intimate Partner Violence Moderates the Association between Oxytocin and Reactivity to Dyadic Conflict among Couples," *Psychiatry Research* 270 (2018): 404–11. **Couples would have to be screened for**: That said, it would be premature to conclude that oxytocin should never be administered in such "problematic" cases. Rather, the existence of possible risk factors for certain individuals suggests that greater initial effort may be required to determine what sort of psychotherapeutic contexts and treatment paradigms would be needed to ensure that such administration resulted in net benefits rather than net harms. In other words, the possibility that oxytocin could prove advantageous (on balance) even for individuals with "risky" profiles—by, for example, facilitating the development of healthier relationship schemas with the guidance of a counselor or therapist—should not be ruled out in advance of attempting the necessary research. This note is adapted from Wudarczyk, Earp, Guastella, and Savulescu, "Could Intranasal Oxytocin Be Used to Enhance Relationships?" **Guastella tells us that data analysis:** Luciana Gravotta, "Be Mine Forever: Oxytocin May Help Build Long-Lasting Love," *Scientific American*, February 12, 2013, www.scientificamerican.com/article/be-mine-forever-oxytocin/. To be clear, the quote is from the *Scientific American* article; however, we also checked with Adam Guastella more recently, and he confirmed that the data have not yet been analyzed; pers. comm., April 12, 2017. **Based on previous findings, the hope:** Gravotta, "Be Mine Forever . . . " **"Oxytocin can elicit loving behaviors":** Quoted in Gravotta, "Be Mine Forever . . . " **combining talk therapy with OT could be:** Gravotta, "Be Mine Forever . . . " **OT can apparently increase:** Michael Kosfled, Markus Heinrichs, Paul J. Zak, Urs Fischbacher, and Ersnt Fehr, "Oxytocin Increases Trust in Humans," *Nature* 435, no. 7042 (2005): 673–76. **It can also increase cooperation:** *Ibid.* **other drugs that affect the OT system, like MDMA:** G. J. H. Dumont, Fred C. G. J. Sweep, R. Van der Steen, Cornelus C. Hermsen, Rogier Donders, Daan J. Touw, Joop M. Van Gerven, Jan Buitelaar, and R. J. Verkes, "Increased Oxytocin Concentrations and Prosocial Feelings in Humans after Ecstasy (3, 4–Methylenedioxymethamphetamine) Administration," *Social Neuroscience* 4, no. 4 (2009): 359–66. **Some bioethicists argue:** Rebecca Bamford, "Unrequited: Neurochemical Enhancement of Love," *Cambridge Quarterly of Healthcare Ethics* 24, no. 3 (2015): 355–60, 358. **In real life, people make their decisions about therapy:** Susan Dodds, "Choice and Control in Feminist Bioethics," in *Relational Autonomy: Feminist Perspectives on Autonomy, Agency and the Social Self*, ed. Catriona Mackenzie and Natalie Stoljar (Oxford: Oxford University Press, 2000), 213–35.

Chapter 9: Anti-love Drugs

References crop up in the writings: William Fitzgerald, "Lucretius' Cure for Love in the 'De Rerum Natura,'" *The Classical World* 78, no. 2 (1984): 73–86. **Ovid:** J. Lewis May, trans., *The Love Books of Ovid: The*

Amores, Ars Amatoria, Remedia Amoris, and Medicamina Faciei Femineae of Publius Ovidius Naso (Whitefish, MT: Kessinger, 2010). **Shakespeare:** William Shakespeare, *As You Like It* (London: Methuen, 1623/1975). **The playwright George Bernard Shaw:** George Bernard Shaw, *The Doctor's Dilemma: Getting Married, and The Shewing-Up of Blanco Posnet* (Edinburgh: Constable & Co., 1911), 34. **Ancient cures for love:** Lawrence Babb, "The Physiological Conception of Love in the Elizabethan and Early Stuart Drama," *Publications of the Modern Language Association of America* 56, no. 4 (1941): 1020–35. **Or if *Harry Potter* is more your style:** As shown in the video game adaptation of J. K. Rowling, *Harry Potter and the Half Blood Prince* (London: Bloomsbury, 2005). See http://harrypotter.wikia.com/wiki/Love_Potion_Antidote. **The following year, writing in:** Larry J. Young, "Being Human: Love: Neuroscience Reveals All," *Nature* 457, no. 7226 (2009): 148. **As Young sees it, love is:** Young, "Being Human . . . ," 148. **He argues that "drugs that:** Young, "Being Human . . . ," 148, emphasis added. **Rabbis and marriage counselors:** Yair Ettinger, "Rabbi's Little Helper," *Haaretz*, April 6, 2012, www.haaretz.com/1.5212045. **In another example, a Christian man:** J. Michael Bostwick and Jeffrey A. Bucci, "Internet Sex Addiction Treated With Naltrexone," *Mayo Clinic Proceedings* 83, no. 2 (2008): 226–30. **And American sex offenders:** Louis J. Gooren, "Ethical and Medical Considerations of Androgen Deprivation Treatment of Sex Offenders," *Journal of Clinical Endocrinology and Metabolism* 96, no. 12 (2011): 3628–37. ***Eternal Sunshine of the Spotless Mind*:** For a thorough and very interesting discussion of the ethics of memory modification as dealt with in this film, see Christopher Grau, "*Eternal Sunshine of the Spotless Mind* and the Morality of Memory," *Journal of Aesthetics and Art Criticism* 64, no. 1 (2006): 119–33. We don't think we'd be able to improve on his analysis, so we aren't going to try. Basically, Grau argues that there are serious costs and other moral issues at stake in artificially manipulating one's memories, but that in some cases it could be still justified. While the possibility of memory erasure is clearly relevant to our general theme in this chapter, a full analysis would take us too far afield of our focus on lust, attraction, and attachment—that is, the predisposing biological factors for potential (harmful) relationships, as well as the psychological glue holding together current (harmful) relationships. The only thing we'll say here is that if a person had already "cured" themself of their emotional attachment to, for example, an abusive partner (perhaps by using one of the interventions we explore in the book), there might be less of a perceived need to delete the partner from their memory as well. For readers who are interested in the science and ethics of memory modification, in addition to the discussion by Grau we recommend the following essays: Reinoudde De Jongh, Ineke Bolt, Maartje Schermer, and Berend Olivier, "Botox for the Brain: Enhancement of Cognition, Mood and Pro-social Behavior and Blunting of Unwanted Memories," *Neuroscience and Biobehavioral Reviews* 32, no. 4 (2008): 760–76; Adam J. Kolber, "Therapeutic Forgetting: The Legal and Ethical Implications of Memory Dampening," *Vanderbilt Law Review* 59 (2006): 1561–2053; and Erik Parens, "The Ethics of Memory Blunting and the Narcissism of Small Differences," *Neuroethics* 3, no. 2 (2010): 99–107. Please note that this note and much of the present chapter is adapted with permission from Brian D. Earp, Olga A. Wudarczyk, Anders Sandberg, and Julian Savulescu, "If I Could Just Stop Loving You: Anti-love Biotechnology and the Ethics of a Chemical Breakup," *American Journal of Bioethics* 13, no. 11 (2013): 3–17. **As Gondry recounts:** Quoted in Jessica Lack, "Eraserhead," *The Guardian*, September 5, 2008, https://www.theguardian.com/film/2008/sep/06/1. **reports of a real-life technique:** Pauline Bock, "How a PTSD Expert Developed a Viable Cure for Heartbreak," *WIRED*, June 26, 2019, https://www.wired.co.uk/article/ptsd-and-heartbreak. All quoted material in the remainder of this section of the chapter is from this source. **More than 70 percent of participants in a 2018 study:** This statistic comes from the *WIRED* article cited in the previous note; however, it is unclear whether the study has been published or whether the author was sharing preliminary results. We did find a 2018 study by Brunet and others that describes a similar technique, but found no mention of romantic betrayal; rather, the patients are simply characterized as having PTSD. See Alain Brunet, Daniel Saumier, Aihua Liu, David L. Streiner, Jacques Tremblay, and Roger K. Pitman, "Reduction of PTSD Symptoms with Pre-reactivation Propranolol Therapy: a Randomized Controlled Trial," *American Journal of Psychiatry* 175, no. 5 (May 2018): 427–33. **As Helen Fisher and her colleagues:** Helen E. Fisher, Debra Mashek, Arthur Aron, Haifang Li, and Lucy L. Brown, "Defining the

Brain Systems of Lust, Romantic Attraction, and Attachment," *Archives of Sexual Behavior* 31, no. 5 (2002): 413–19. **For example, testosterone can stimulate:** See Helen Fisher, "Lust, Attraction, Attachment: Biology and Evolution of the Three Primary Emotion Systems for Mating, Reproduction, and Parenting," *Journal of Sex Education and Therapy* 25, no.1 (2002): 96–104. **sophisticated nanobiochips and advances in brain imaging will allow:** Zack Lynch, "Neurotechnology and Society (2010–2060)," *Annals of the New York Academy of Sciences* 1013, no. 1 (2004): 229–33. **We can also add a few household examples:** David M. Nudell, Mara M. Monoski, and Larry I. Lipshultz, "Common Medications and Drugs: How They Affect Male Fertility," *The Urologic Clinics of North America* 29, no. 4 (2002): 965–73. **These include almost all blood:** Howard LeWine, "Sexual Side Effects from Blood Pressure Medicine Can Affect Both Men and Women," *The Medicine Cabinet*, August 2, 2017, https://tribunecontentagency.com/article/sexual-side-effects-from-blood-pressure-medicine-can-affect-both-men-and-women/. **pain relievers containing butalbital**: Xiulu Ruan, "Drug-Related Side Effects of Long-Term Intrathecal Morphine Therapy," *Pain Physician* 10, no. 2 (2007): 357. **statin cholesterol drugs:** Laurens De Graaf et al., "Is Decreased Libido Associated with the Use of HMG-CoA-reductase Inhibitors?" *British Journal of Clinical Pharmacology* 58, no. 3 (2004): 326–28. **certain acid blockers:** Armon B. Neel Jr., "7 Meds That Can Wreck Your Sex Life," AARP, 2012, www.aarp.org/health/drugs-supplements/info-04-2012/medications-that-can-cause-sexual-dysfunction.html. **the hair loss drug finasteride:** Michael S. Irwig, and Swapna Kolukula, "Persistent Sexual Side Effects of Finasteride for Male Pattern Hair Loss," *Journal of Sexual Medicine* 8, no. 6 (2011): 1747–53. **and seizure medications:** Arthur C. Grant and Hyunjue Oh, "Gabapentin-Induced Anorgasmia in Women," *American Journal of Psychiatry* 159, no. 7 (2002): 1247. **Focusing on this hormone as one of the most important:** The relationship between testosterone and libido is not as straightforward as popular thinking suggests, but testosterone does play at least some important roles in the biology of human sexual desire. For evidence and discussions of the (complicated) link between testosterone and libido, see, for example, Thomas G. Travison, John E. Morley, Andre B. Araujo, Amy B. O'Donnell, and John B. McKinlay, "The Relationship Between

Libido and Testosterone Levels in Aging Men," *The Journal of Clinical Endocrinology & Metabolism* 91, no. 7 (2006): 2509–2513; Giulia Rastrelli, Giovanni Corona, and Mario Maggi, "Testosterone and Sexual Function in Men," *Maturitas* 112 (2018): 46–52; Katrina Karkazis, "The Masculine Mystique of T," *The New York Review of Books*, June 28, 2018, https://www.nybooks.com/daily/2018/06/28/the-masculine-mystique-of-t/. **One study, for example, reported that cutting:** Ariel Rösler and Eliezer Witztum, "Treatment of Men with Paraphilia with a Long-Acting Analogue of Gonadotropin-Releasing Hormone," *New England Journal of Medicine* 338, no. 7 (1998): 416–22. **Likewise, the neuroscientist Till Amelung:** Till Amelung, Laura F. Kuhle, Anna Konrad, Alfred Pauls, and Klaus M. Beier, "Androgen Deprivation Therapy of Self-Identifying, Help-Seeking Pedophiles in the Dunkelfeld," *International Journal of Law and Psychiatry* 35, no. 3 (2012): 176–84. **These included pedophilia, voyeurism:** Richard B. Kruger and Meg S. Kaplan, "Depot-Leuprolide Acetate for Treatment of Paraphilias: A Report of Twelve Cases," *Archives of Sexual Behavior* 30, no. 4 (2001): 409–22. The other quotes in this paragraph are also from this source. **Around that time, the Finnish anthropologist:** Edward Westermarck, *The History of Human Marriage* (New York: Macmillan, 1921). **Although the exact mechanism underlying:** Mark A. Schneider and Lewellyn Hendrix, "Olfactory Sexual Inhibition and the Westermarck Effect," *Human Nature* 11, no. 1 (2000): 65–91; Glenn E. Weisfeld, Tiffany Czilli, Krista A. Phillips, James A. Gall, and Cary M. Lichtman, "Possible Olfaction-Based Mechanisms in Human Kin Recognition and Inbreeding Avoidance," *Journal of Experimental Child Psychology* 85, no. 3 (2003): 279–95. **Whatever the mechanism, there seems:** Liqun Luo, "Is There a Sensitive Period in Human Incest Avoidance?" *Evolutionary Psychology* 9, no. 2 (2011): 285–95. **which raises the intriguing possibility:** Takao K. Hensch and Parizad M. Bilimoria, "Re-opening Windows: Manipulating Critical Periods for Brain Development," in *Cerebrum: The Dana Forum on Brain Science* (Dana Foundation, 2012). **The idea was that the "same lack:** As summarized in Kayt Sukel, *Dirty Minds: How Our Brains Influence Love, Sex, and Relationships* (New York: Free Press, 2012), 35. **As Marazziti and her coauthors conclude:** Donatella Marazziti, Hagop S. Akiskal, Alessandro Rossi, and Giovanni B. Cassano, "Alteration of the Platelet

Serotonin Transporter in Romantic Love," *Psychological Medicine* 29, no. 3 (1999): 741–45. **Indeed, retesting the lovers:** Marazziti et al., "Alteration of the Platelet Serotonin Transporter . . . ," 744. **As it happens, patients with OCD:** See, e.g., Stuart A. Montgomery, "Long-Term Treatment of Depression," *British Journal of Psychiatry* 165, no. S26 (1994): 31–36; and Joseph Zohar and Thomas R. Insel, "Drug Treatment of Obsessive-Compulsive Disorder," *Journal of Affective Disorders* 13, no. 2 (1987): 193–202. **80 percent of SSRI-using patients:** Adam Opbroek as quoted in Tara Parker-Pope, "Where Is the Love? Antidepressants May Inadvertently Blunt Feelings of Romance," *Wall Street Journal*, February 14, 2006, www.wsj.com/articles/SB113987710213672933. **Again, if you're trying to maintain:** We are not suggesting anyone take SSRIs to help pull off a breakup. For one thing, you would need a prescription from your doctor, and saying "I hope the drug will help me care less about my partner's feelings" is not going to cut it. We are just talking about effects and mechanisms right now, not the ethics of off-label drug use. **There is, however, compelling evidence:** See, e.g., Thomas R. Insel and Larry J. Young, "The Neurobiology of Attachment," *Nature Reviews Neuroscience* 2, no. 2 (2001): 129–36; and Thomas R. Insel, Larry J. Young, and Zuoxin Wang, "Central Oxytocin and Reproductive Behaviours," *Reviews of Reproduction* 2, no. 1 (1997): 28–37. **In one study, injecting female prairie voles:** Y. Liu and Z. X. Wang, "Nucleus Accumbens Oxytocin and Dopamine Interact to Regulate Pair Bond Formation in Female Prairie Voles," *Neuroscience* 121, no. 3 (2003): 537–44. **As Larry Young put it:** Quoted in John Tierny, "A Love Vaccine?" *New York Times*, January 12, 2009, https://tierneylab.blogs.nytimes.com/2009/01/12/love-vaccine/. **Likewise, pair-bonded male prairie voles:** Brandon J. Aragona, Yan Liu, Y. Joy Yu, J. Thomas Curtis, Jacqueline M. Detwiler, Thomas R. Insel, and Zuoxin Wang, "Nucleus Accumbens Dopamine Differentially Mediates the Formation and Maintenance of Monogamous Pair Bonds," *Nature Neuroscience* 9, no. 1 (2006): 133–39. **As we mentioned before, most scientists:** See Larry J. Young, "The Neural Basis of Pair Bonding in a Monogamous Species: A Model for Understanding the Biological Basis of Human Behavior," in *Offspring: Human Fertility Behavior in Biodemographic Perspective*, ed. Kenneth W. Wachter and Rodolfo A. Bulato (Washington, DC: National Academies

Press, 2003), 91–103. **Consider a recent headline from VICE magazine:** Sirin Kale, "How to Bio-Hack Your Brain to Have Sex Without Getting Emotionally Attached," *VICE*, August 25, 2016, https://www.vice.com/en_us/article/59mmzq/how-to-bio-hack-your-brain-to-have-sex-without-getting-emotionally-attached. The quotes from Larry Young in the following paragraphs are also from this source. **Research shows that prolonged eye contact:** Even between humans and dogs! See Miho Nagasawa, Shouhei Mitsui, Shiori En, Nobuyo Ohtani, Mitsuaki Ohta, Yasuo Sakuma, Tatsushi Onaka, Kazutaka Mogi, and Takefumi Kikusui, "Oxytocin-gaze Positive Loop and the Coevolution of Human-Dog Bonds," *Science* 348, no. 6232 (2015): 333–36. **Young and his collaborators allowed male prairie voles:** Allison Anacker, Todd H. Ahern, Caroline M. Hostetler, Brett D. Dufour, Monique L. Smith, Davelle L. Cocking, Ju Li, Larry J. Young, Jennifer M. Loftis, and Andrey E. Ryabinin, "Drinking Alcohol has Sex-dependent Effects on Pair Bond Formation in Prairie Voles," *Proceedings of the National Academy of Sciences* 111, no. 16 (2014): 6052–57. **When given the choice to huddle:** Allison Anacker et al., "Drinking Alcohol," 6052. **Another, more speculative lead:** J. Capgras and J. Reboul-Lachaux, "L'illusion des 'Sosies' dans un délire systématisé chronique," *History of Psychiatry* 5, no. 17 (1994): 119–33. **Patients suffering from this condition:** Hadyn D. Ellis and Andrew W. Young, "Accounting for Delusional Misidentifications," *British Journal of Psychiatry* 157, no. 2 (1990): 239–48. **One explanation for this phenomenon:** Hadyn D. Ellis, Andrew W. Young, Angela H. Quayle, and Karel W. De Pauw, "Reduced Autonomic Responses to Faces in Capgras Delusion," *Proceedings of the Royal Society B: Biological Sciences* 264, no. 1384 (1997): 1085–92. **One question this raises, then:** We thank John Danaher for pushing us on this. John Danaher, "The Vice of In-Principlism and the Harmfulness of Love," *American Journal of Bioethics* 13, no. 11 (2013): 19–21. **Sometimes the evidence concerning appropriate:** Madlen Gazarian, Maria Kelly, John R. McPhee, Linda V. Graudins, Robyn L. Ward, and Terence J. Campbell, "Off-Label Use of Medicines: Consensus Recommendations for Evaluating Appropriateness," *Medical Journal of Australia* 185, no. 10 (2006): 544–48. **people are afraid of pathologizing love:** Diana Aurenque and Christopher W. McDougall, "Amantes Sunt Amentes: Pathologizing Love and the Meaning of

Suffering," *The American Journal of Bioethics* 13, no. 11 (2013): 34–36. **we should be open to the idea:** Matthis Synofzik, "Ethically Justified, Clinically Applicable Criteria for Physician Decision-making in Psychopharmacological Enhancement," *Neuroethics* 2, no. 2 (2009): 89–102.

Chapter 10: Chemical Breakups

After a trifling incident early: Quoted in Susan McClelland, "When Love Hurts: The Story of an Abused Woman," *Canadian Living*, October 27, 2006, www.canadianliving.com/life-and-relationships/relationships/article/when-love-hurts-the-story-of-an-abused-woman. All subsequent quotes from Bonnie are from this interview. **Bonnie's experience of what she:** When we first submitted the paper this chapter is based on, a reviewer wrote: "This is Bonnie's experience—how she conceptualizes what she feels. However, a psychiatrist or psychologist may say that what she feels is not actually love but an obsessive attachment to the abuser, or an emotional allegiance from the perspective of a criminologist, [or] a mental disorder." This is undoubtedly true: one person's "love" may certainly be thought of as "insanity" by someone else—or a delusion, or none of the above. But we want to be careful about deciding for other people what their own experience should properly be called, as we explained in Chapter 1. A psychiatrist, for example, might certainly want to define "true" love as being something intrinsically healthy, positive, and overall good for one's well-being; on that definition, we would have to conclude that Bonnie was mistaken about her own feelings or was using the word "love" incorrectly. Yet other definitions abound. The philosopher Simon May, for example, discusses conceptions of "true" love according to which it may sometimes be destructive, even to the point of death: Simon May, *Love: A History* (New Haven: Yale University Press, 2011). And Sarah E. Taylor gives a very thoughtful phenomenological analysis of "loving the ones who are violent to us," questioning the predominant view among researchers that "love and violence are dichotomous, or that love cannot remain in spite of violence [such that] any individual who experiences love for a violent other is somehow pathological or compromised." See Sarah E. Taylor, "Loving the Ones who are Violent to Us: An Existential Phenomenological Study," doctoral dissertation, The Chicago School of Professional Psychology,

2017. Yet whatever position one takes on the question of labeling, the moral analysis remains similar. Are the feelings harmful? Why? By virtue of what? And how might they best be tempered or resolved? This note, and much of the current chapter, is adapted with permission from Brian D. Earp, Olga A. Wudarczyk, Anders Sandberg, and Julian Savulescu, "If I Could Just Stop Loving You: Anti-love Biotechnology and the Ethics of a Chemical Breakup," *American Journal of Bioethics* 13, no. 11 (2013): 3–17. **Such cases may even represent:** Jospeh M. Carver, *Love and Stockholm Syndrome: The Mystery of Loving an Abuser* (Kenmore: Mental Health Matters, 2007). **Harmful relationships can be literally:** Stanton Peele and Archie Brodsky, "Love Can Be an Addiction," *Psychology Today*, 1974, 22–26; Stanton Peele and Archie Brodsky, *Love and Addiction* (New York: Broadrow, 1975). See also Brian D. Earp, Olga A. Wudarczyk, Bennett Foddy, and Julian Savulescu, "Addicted to Love: What Is Love Addiction and When Should It Be Treated," *Philosophy, Psychiatry, and Psychology* 24, no.1 (2017): 77–92. **"About 10 months ago:** TEENADVOCATEDAN, "How Do I Stop Loving the Abuser," *Feministing,* 2009, http://feministing.com/2009/10/20/how-do-i-stop-loving-the-abuser/. Please note that the quote has been slightly edited for readability. **(unrequited love):** For a discussion of the ethics of "treating" unrequited love, see Francesca Minerva, "Unrequited Love Hurts," *Cambridge Quarterly of Healthcare Ethics* 24, no. 4 (2015): 479–85. **(erotomania):** See J. Reid Meloy, "Unrequited Love and the Wish to Kill: Diagnosis and Treatment of Borderline Erotomania," *Bulletin of the Menninger Clinic* 53, no. 6 (1989): 477–92. **voluntary BDSM relationships:** BDSM stands for bondage/discipline, domination/submission, sadism/masochism. For a thoughtful discussion of popular understandings of BDSM, see Meg Barker, "Consent Is a Grey Area? A Comparison of Understandings of Consent in 'Fifty Shades of Grey' and on the BDSM Blogosphere," *Sexualities* 16, no.8 (2013): 896–914. See also Rebecca Kukla, "That's What She Said: The Language of Sexual Negotiation," *Ethics* 129, no. 1 (2018): 70–97. **It is a history of coercively applying:** Thomas Szasz, *Coercion as Cure: A Critical History of Psychiatry* (Piscataway, NJ: Transaction, 2009). **The primary goal then should be:** We thank Diana Aurenque and Christopher McDougall for highlighting this issue. See Diana Aurenque and Christopher W. McDougall,

"Amantes Sunt Amantes: Pathologizing Love and the Meaning of Suffering," *American Journal of Bioethics* 13, no. 11 (2013): 34–36, 35. **"Life became hell:** Quoted in McClelland, *When Love Hurts* . . . **First, many people who are in abusive relationships:** Silke Meyer, "Why Women Stay: A Theoretical Examination of Rational Choice and Moral Reasoning in the Context of Intimate Partner Violence," *Australian & New Zealand Journal of Criminology* 45, no. 2 (2012): 179–193. **We do not wish to blame the victim:** Again, we are grateful to Diana Aurenque and Christopher McDougall for raising the concern that an earlier version of our argument could be misinterpreted that way. We must also credit them for the important point that some individuals may fear putting their children in danger by leaving an abusive relationship. **Although the case of pedophiles:** Aurenque and McDougall, "Amantes Sunt Amentes . . . ," 35–36. **Because of the stigmatization:** Sara Jahnke and Juergen Hoyer, "Stigmatization of People with Pedophilia: A Blind Spot in Stigma Research," *International Journal of Sexual Health* 25, no. 3 (2013): 169–184. **You could dwell on all the ways:** As Ovid advised long ago in his *Remedia amoris*: "Tell yourself often what your wicked girl has done, and before your eyes place every hurt you've had. Impress your mind with whatever's wrong with her body, and keep your eyes fixed all the time on those faults." See J. Lewis May, trans., *The Love Books of Ovid: the Amores, Ars Amatoria, Remedia Amoris, and Medicamina Faciei Femineae of Publius Ovidius Naso* (Whitefish, MT: Kessinger, 2010). **"A skin contusion or broken bone:** Neil McArthur, "The Heart Outright: A Comment on 'If I Could Just Stop Loving You,'" *American Journal of Bioethics* 13, no. 11 (2013): 24–25, 24. The subsequent quotes in this paragraph are from the same source. **"I think, in a way," he said:** Ben Sessa, pers. comm., September 27, 2017. Quoted with consent. **"With suffering comes understanding":** Erik Parens, "On Good and Bad Forms of Medicalization," *Bioethics* 27, no. 1 (2013): 28–35, 32. **pharmacological and traditional methods "can both achieve":** Parens, "On Good and Bad Forms . . . ," emphasis added. **"We should be slower to imagine:** Parens, "On Good and Bad Forms . . . ," 32. **But even if some drugs do introduce:** Christopher Grau, "*Eternal Sunshine of the Spotless Mind* and the Morality of Memory," *Journal of Aesthetics and Art Criticism* 64, no. 1 (2006): 119–33, 133. **Here, then, is a summary:**

In the original paper on which this chapter is primarily based, we had four conditions, but on reflection and in response to various criticisms, we have decided that we prefer these three. **policymakers, doctors, and individuals:** McArthur, "Heart Outright . . . ," 24.

Chapter 11: Avoiding Disaster

Instead, the harms that might come: Nick Bostrom and Rebecca Roache, "Ethical Issues in Human Enhancement," in *New Waves in Applied Ethics*, ed. Jesper Ryberg (New York: Palgrave Macmillan, 2008), 120–52. **The basic idea is that some:** Caroline M. Parker, Jennifer S. Hirsch, Morgan M. Philbin, and Richard G. Parker, "The Urgent Need for Research and Interventions to Address Family-Based Stigma and Discrimination against Lesbian, Gay, Bisexual, Transgender, and Queer Youth," *Journal of Adolescent Health* 63, no. 4 (2018): 383–93. **If we are going to get on board with:** Kristina Gupta, "Protecting Sexual Diversity: Rethinking the Use of Neurotechnological Interventions to Alter Sexuality," *AJOB Neuroscience* 3, no. 3 (2012): 24–28. **merely having the option to change:** For an extensive recent discussion of the ways in which having certain options can be harmful, see Nathan Hodson, Lynne Townley, and Brian D. Earp, "Removing Harmful Options: The Law and Ethics of International Commercial Surrogacy," *Medical Law Review*, in press. **In essence, they would be forced:** For an excellent discussion of this kind of worry, see Candice Delmas and Sean Aas, "Sexual Reorientation in Ideal and Non-ideal Theory," *Journal of Political Philosophy* 26, no. 4 (2018): 463–85. **The practice of so-called conversion therapy:** Timothy F. Murphy, "Redirecting Sexual Orientation: Techniques and Justifications," *Journal of Sex Research* 29, no. 4 (1992): 501–23. **And as late as 2012:** Evan Halper, "Judge Blocks Ban on Gay 'Conversion' Therapy," *Los Angeles Times*, December 3, 2012, http://latimesblogs.latimes.com/california-politics/2012/12/judge-blocks-ban-on-gay-conversion-therapy.html. Please note that since this ruling, some individual states have in fact passed bans or instituted strict regulations concerning sexual orientation change efforts: Susan Miller, "Record Number of States Banning Conversion Therapy," *USA Today*, April 17, 2018, www.usatoday.com/story/news/nation/2018/04/17/states-banning-conversion-therapy/518972002/. **It is still being performed:** Julie Moreau, "Thousands of

Teens Will Undergo 'Conversion Therapy' in Near Future, Study Estimates," *NBC News*, January 26, 2018, www.nbcnews.com/feature/nbc-out/80-000-teens-will-undergo-conversion-therapy-near-future-study-n841356. **Historical efforts to modify:** David B. Cruz, "Controlling Desires: Sexual Orientation Conversion and the Limits of Knowledge and Law," *Southern California Law Review* 72 (1998): 1297–1400; Douglas C. Haldeman, "Gay Rights, Patient Rights: The Implications of Sexual Orientation Conversion Therapy," *Professional Psychology Research and Practice* 33, no. 3 (2002): 260–64; Charles E. Moan and Robert G. Heath, "Septal Stimulation for the Initiation of Heterosexual Behavior in a Homosexual Male," *Journal of Behavior Therapy and Experimental Psychiatry* 3, no. 1 (1972): 23IN127–2630; Udo Schüklenk, Edward Stein, Jacinta Kerin, and William Byne, "The Ethics of Genetic Research on Sexual Orientation," *Hastings Center Report* 27, no. 4 (1997): 6–13. Please note that this paragraph and some other portions of this chapter are adapted with permission from Brian D. Earp, Anders Sandberg, and Julian Savulescu, "Brave New Love: The Threat of High-Tech 'Conversion' Therapy and the Bio-oppression of Sexual Minorities," *AJOB Neuroscience* 5, no. 1 (2014): 4–12. **Psychiatric drugs are being given to Orthodox:** Yair Ettinger, "Rabbi's Little Helper," *Haaretz*, April 6, 2012, www.haaretz.com/1.5212045. For a related discussion, see Batya Ungar-Sargon, "Healing Hasidic Adulterers with Psychiatric Drugs," *The Establishment*, September 7, 2016, https://medium.com/the-establishment/healing-hasidic-adulterers-with-psychiatric-drugs-8e4fa663c035. **In the United States in 2015:** Valerie Jarrett, "Petition Response: On Conversion Therapy," The White House, April 8, 2015, https://obamawhitehouse.archives.gov/blog/2015/04/08/petition-response-conversion-therapy. Please note that this portion and some other sections of the chapter are adapted with permission from Brian D. Earp and Andrew Vierra, "Sexual Orientation Minority Rights and High-Tech Conversion Therapy," in *Handbook of Philosophy and Public Policy*, ed. D. Boonin (Basingstoke, UK: Palgrave Macmillan, 2018), 535–50. Additional material is adapted from Andrew Vierra and Brian D. Earp, "Born This Way? How High-Tech Conversion Therapy Could Undermine Gay Rights," *The Conversation*, April 21, 2015, https://theconversation.com/born-this-way-how-high-tech-conversion-therapy-could-undermine-gay-rights-40121.

The Human Rights Campaign made: "The Lies and Dangers of Efforts to Change Sexual Orientation or Gender Identity," Human Rights Campaign, www.hrc.org/resources/the-lies-and-dangers-of-reparative-therapy. **Specifically, the physical, mental:** "Report of the Task Force on Appropriate Therapeutic Responses to Sexual Orientation," American Psychological Association, 2009, www.apa.org/pi/lgbt/resources/therapeutic-response.pdf. **And apart from a smattering of:** Robert L. Spitzer, "Can Some Gay Men and Lesbians Change Their Sexual Orientation? 200 Participants Reporting a Change from Homosexual to Heterosexual Orientation," *Archives of Sexual Behavior* 32, no. 5 (2003): 403–17; Jack Drescher, "Can Sexual Orientation Be Changed?" *Journal of Gay and Lesbian Mental Health* 19, no. 1 (2015): 84–93; American Psychological Association, "Report of the Task Force . . . " **Could such efforts be morally permissible:** For a classic discussion of these questions, see Timothy F. Murphy, *Gay Science: The Ethics of Sexual Orientation Research* (New York: Columbia University Press, 1997). **Based on current trends in research:** See Earp, Sandberg, and Savulescu, "Brave New Love . . . " **(For example, what is the sexual orientation of:** Robin A. Dembroff, "What Is Sexual Orientation?" *Philosophers' Imprint* 16, no. 3 (2016): 1–27. **"Same Love":** Featuring Mary Lambert. See https://genius.com/Macklemore-and-ryan-lewis-same-love-lyrics. For an in-depth critique of the argument from immutability, see Janet E. Halley, "Sexual Orientation and the Politics of Biology: A Critique of the Argument from Immutability," *Stanford Law Review* 46, no. 3 (1994): 503–68. **and now even "transracial" identities are being discussed:** Rebecca Tuvel, "In Defense of Transracialism," *Hypatia* 32, no. 2 (2017): 263–78; Tina Botts, "Race and Method: The Tuvel Affair," *Philosophy Today* 62, no. 1 (2018): 51–72; Rogers Brubaker, *Trans: Gender and Race in an Age of Unsettled Identities* (Princeton: Princeton University Press, 2016); Cressida J. Heyes, "Changing Race, Changing Sex: The Ethics of Self-Transformation," *Journal of Social Philosophy* 37, no. 2 (2006): 266–82; Christine Overall, "Transsexualism and 'Transracialism,'" *Social Philosophy Today* 20 (2004): 183–93; Jimmie Manning and Jennifer C. Dunn, "Rachel Dolezal, Transracialism, and the Hypatia Controversy: Difficult Conversations and the Need for Transgressing Feminist Discourses," in *Transgressing Feminist Theory and Discourse*, ed. Jimmie Manning and Jennifer C.

Dunn (Abingdon, UK: Routledge, 2018), 21–33; Kris Sealey, "Transracialism and White Allyship: A Response to Rebecca Tuvel," *Philosophy Today* 62, no. 1 (2018): 21–29. Please note that with respect to transitions in identity based on gender this does not necessarily describe the experience of many transgender individuals, who state that they are discovering or manifesting a gender they have always had (sometimes supplemented by changes in bodily sex characteristics), rather than that they are *changing* from one gender to another. However, some transgender and gender-fluid individuals do feel that their gender identity or gender experience can and does change through time. These issues should be kept conceptually distinct. **"The timeline of events in history:** Mark Bailey, quoted in Brian D. Earp, "Choosing One's Own (Sexual) Identity: Shifting the Terms of the 'Gay Rights' Debate," *Practical Ethics*, January 26, 2012, http://blog.practicalethics.ox.ac.uk/2012/01/can-you-be-gay-by-choice/. **Just like people with heterosexual desires:** Simon LeVay, *Gay, Straight, and the Reason Why: The Science of Sexual Orientation* (Oxford: Oxford University Press, 2016). **All of us deserve to live, love:** Paraphrased from Tia Powell, Sophia Shapiro, and Ed Stein, "Transgender Rights as Human Rights," *AMA Journal of Ethics* 18, no. 11 (2016): 1127–31, 1129. **Can you choose to be gay?:** This portion of the chapter is adapted from Brian D. Earp, "Can You Be Gay by Choice?" in *Philosophers Take on the World*, ed. David Edmonds (Oxford: Oxford University Press, 2016), 95–98. **"I gave a speech recently:** Cynthia Nixon, quoted in Alex Witchell, "Life after 'Sex,'" *New York Times Magazine*, January 19, 2012, www.nytimes.com/2012/01/22/magazine/cynthia-nixon-wit.html. All quotes from Nixon are from this source. **No one can get inside your head:** The following paragraph is adapted from Earp, "Choosing One's Own (Sexual) Identity . . . " **neither sex nor gender are simple binaries:** For a nice introduction, see Claire Ainsworth, "Sex Redefined," *Nature* 518 (2015): 288–91; see also Diana Elena Moga, "So What Is Gender Anyway? And Who's Having Sex with Whom?" *Journal of the American Psychoanalytic Association* 66, no. 3 (2018): 527–43. **our more basic sexual orientations are, as far as scientists can tell:** For an overview and general discussion, see J. Michael Bailey, Paul L. Vasey, Lisa M. Diamond, S. Marc Breedlove, Eric Vilain, and Marc Epprecht, "Sexual Orientation, Controversy, and Science," *Psychological Science in* *the Public Interest* 17, no. 2 (2016): 45–101. **(As the mayor of South Bend, Indiana:** Quoted in Katie Reilly, "Pete Buttigieg Criticizes Vice President Pence in Speech About LGBTQ Rights: 'Your Quarrel, Sir, Is With My Creator,'" *TIME*, April 9, 2019, https://time.com/5566322/pete-buttigieg-mike-pence-lgbtq-speech/. **But even Cynthia Nixon, if she:** For a critical assessment of this claim, see William S. Wilkerson, "Is It a Choice? Sexual Orientation as Interpretation," *Journal of Social Philosophy* 40, no.1 (2009): 97–116. See also Esa Díaz-León, "Sexual Orientation as Interpretation? Sexual Desires, Concepts, and Choice," *Journal of Social Ontology* 3, no. 2 (2017): 231–48. **"Even if sexual orientation is not chosen:** Tia Powell and Edward Stein, "Legal and Ethical Concerns about Sexual Orientation Change Efforts," *Hastings Center Report* 44, no. s4 (2014): S32–S39, S36–S37. **"Religious conservatives go on TV:** Dan Savage, "Ben Carson: Being Gay Is a Choice and Prison Proves It," *The Stranger*, March 4, 2015, www.thestranger.com/blogs/slog/2015/03/04/21827375/republican-idiot-being-gay-is-a-choice-and-prison-proves-it. All quotes from Savage in this chapter are from this source. **Few human traits are:** Powell and Stein, "Legal and Ethical Concerns," S35. **We will focus on the case of the yeshiva:** Much of the rest of this chapter is adapted from Earp and Vierra, "Sexual Orientation Minority Rights and High-Tech Conversion Therapy . . . " **"Some behaviors put Haredim:** Quoted in Ettinger, "Rabbi's Little Helper . . . " **Second, we believe that what ultimately needs changing:** For some arguments why, concerning an analogous case, see Brian D. Earp, "The Ethics of Infant Male Circumcision," *Journal of Medical Ethics* 39, no. 7 (2013): 418–20; see also Brian D. Earp and Robert Darby, "Circumcision, Sexual Experience, and Harm," *University of Pennsylvania Journal of International Law* 37, No. 2 (online) (2017): 1–56, especially 45–47. **"Homophobic attitudes have been:** Douglas C. Haldeman, "The Practice and Ethics of Sexual Orientation Conversion Therapy," *Journal of Consulting and Clinical Psychology* 62, no. 2 (1994): 221. **Indeed, this is exactly the approach:** Sarah L. Schulz, "The Informed Consent Model of Transgender Care: An Alternative to the Diagnosis of Gender Dysphoria," *Journal of Humanistic Psychology* 58, no. 1 (2018): 72–92. **Primarily, this is because such procedures:** For some recent, related evidence and discussion, see Hillary Nguyen, Alexis Chavez, Emily Lipner, Liisa Hantsoo,

Sara L. Kornfield, Robert D. Davies, and Cynthia Neill Epperson, "Gender-Affirming Hormone Use in Transgender Individuals: Impact on Behavioral Health and Cognition," *Current Psychiatry Reports* 20, no. 110 (2018): 1–9. **"clinicians would often be permitted, and sometimes even required:** Sean Aas and Candice Delmas, "The Ethics of Sexual Reorientation: What Should Clinicians and Researchers Do? *Journal of Medical Ethics* 42, no. 6 (2016): 340–47, 341. **In fact, Delmas and Aas have argued that:** Candice Delmas, "Three Harms of 'Conversion' Therapy," *AJOB Neuroscience* 5, no. 2 (2014): 22–23. See also more recently Delmas and Aas, "Sexual Reorientation. . ." Note that in this paper they discuss a fourth harm, which is that the power of the "born this way" argument for gay rights would be weakened by the availability of HCT, an issue we covered earlier in the chapter. **In this world, where nonheterosexuality is:** Delmas, "Three Harms . . . ," 22. See also, generally, Delmas and Aas, "Sexual Reorientation . . . " **This harm, they suggest, could manifest:** Delmas and Aas, "Sexual Reorientation . . . " **According to Delmas and Aas, such a situation:** Delmas and Aas, "Sexual Reorientation . . . " **Just as bisexual people are sometimes expected:** Delmas and Aas, "Sexual Reorientation . . . " **or even the erasure of their "kind" of person:** Assuming that they identify as members of a human "kind" based on sexual orientation. For a related discussion, see Jason Behrmann and Vardit Ravitsky, "Turning Queer Villages into Ghost Towns: A Community Perspective on Conversion Therapies," *AJOB Neuroscience* 5, no. 1 (2014): 14–16. **Taken together, they think, the harms make:** Delmas, "Three Harms . . . ," 22. **According to the English writer:** Julie Bindel, "My Sexual Revolution," *The Guardian*, January 30, 2009, www.theguardian.com/lifeandstyle/2009/jan/30/women-gayrights. **The book did not insist that women:** Bindel, "My Sexual Revolution . . . " **Given the current situation in which relatively few people:** However, see Rhys Southan, "Re-Orientation," *Medium*, July 8, 2019, https://medium.com/@rhys/re-orientation-fb131ba7bd9b. **This sets us up for a dilemma:** Peter Murphy, "Help the Patient, But Be Complicit with Homophobic Social Norms? Four Issues," *AJOB Neuroscience* 5, no. 1 (2014): 13–14. **We might think that for the sake of:** Felicitas Kraemer, "A Technological Fix for the Self? How Neurotechnologies Shape Who We are and Whom We Love," *AJOB Neuroscience* 5, no. 1 (2014): 1–3.

Anyone accused of being complicit: This analogy is from Thomas Murray, "Enhancement," in *The Oxford Handbook of Bioethics*, ed. Bonnie Steinbock (Oxford: Oxford University Press, 2007), 491–515, 511. **Feminist philosophers such as Margaret Olivia Little:** Margaret Olivia Little, "Cosmetic Surgery, Suspect Norms, and the Ethics of Complicity," in *Enhancing Human Traits: Ethical and Social Implications*, ed. Erik Parens (Washington, DC: Georgetown University Press, 1998), 162–76. **sacrificed on the altar:** This phrase is from the introduction to Erik Parens, ed., *Surgically Shaping Children: Technology, Ethics, and the Pursuit of Normality* (Baltimore: Johns Hopkins University Press, 2006). **Ultimately, Little splits the difference:** As summarized by Thomas Murray, "Enhancement," 512. For an alternative perspective arguing for state prohibition of practices that cause harm to those who ostensibly choose them (assuming that the choice is due only to unjust social pressures), see Clare Chambers, "Are Breast Implants Better than Female Genital Mutilation? Autonomy, Gender Equality and Nussbaum's Political Liberalism," *Critical Review of International Social and Political Philosophy* 7, no. 3 (2004): 1–33, and more broadly, Clare Chambers, *Sex, Culture, and Justice: The Limits of Choice* (University Park, PA: Pennsylvania State University Press, 2008). **"Ideally the individual ultimately integrates:** Douglas Haldeman, "Gay Rights, Patient Rights: The Implications of Sexual Orientation Conversion Therapy," *Professional Psychology Research and Practice* 33, no. 3 (2002): 260–64, 263. **how can we use new technologies for good:** For discussion, see Anders Sandberg, "Cognition Enhancement: Upgrading the Brain," in *Enhancing Human Capacities*, ed. Julian Savulescu, Ruud ter Meulen, and Guy Kahane (Oxford: Wiley-Blackwell, 2011).

Chapter 12: Choosing Love

"Our meddling intellect": William Wordsworth, *The Major Works* (Oxford: Oxford University Press, 2008). **Carrie Jenkins, the philosopher who:** Carrie Jenkins, *What Love Is: And What It Could Be* (New York: Basic Books, 2017), (advance copy). **As Jenkins argues, treating love:** Jenkins, *What Love Is*, 9. **Here is an analogy:** We think this analogy comes from an interview with Helen Fisher, but we can't remember when or where, so we will just give our own version. **Significantly, he concludes with:** Quoted in Erik Parens, "On Good and Bad Forms of

Medicalization," *Bioethics* 27, no. 1 (2013): 28–35, 33. Note that the paper by Evans has not yet been published, so we must rely on Erik Parens's reading of the passage. However, we wish to point out that as quoted, it is unclear whether the claim that "we can all fear the medicalization of love" is one that Evans himself would endorse, or whether, as a sociologist, he is intending only to illustrate the sort of complaint that might be raised by critics of medicalization generally. Thus, our argument should be interpreted as being a response to the claim (and the sorts of worries that we think might be lying behind it), rather than as a rebuttal to Evans. **In our research, we have identified four main worries:** Versions of these worries are discussed at length in Peter Conrad, *The Medicalization of Society: On the Transformation of Human Conditions into Treatable Disorders* (Baltimore: Johns Hopkins University Press, 2008). **Medicalization can transform ordinary:** Conrad, *Medicalization of Society*, 148. **Medicalization can expand the influence:** David Armstrong, "The Rise of Surveillance Medicine," *Sociology of Health and Illness* 17, no. 3 (1995): 393–404. **It can also create openings for:** Conrad, *Medicalization of Society*, 49–151. **Medicalization can reframe social problems:** Conrad, *Medicalization of Society*, 152. See also Barbara Wootton, *Social Science and Social Pathology* (New York: Macmillan, 1959). **This can lead to an undue emphasis on:** Erik Parens, "Toward a More Fruitful Debate about Enhancement," in *Human Enhancement*, ed. Julian Savulescu and Nick Bostrom (Oxford, UK: Oxford University Press, 2009), 184. **Unwanted pregnancy, for example, is not a disease:** Parens, "On Good and Bad Forms of Medicalization," 33. **Hypoactive sexual desire disorder appears to be:** The account in this section draws heavily from Antonie Meixel, Elena Yanchar, and Adriane Fugh-Berman, "Hypoactive Sexual Desire Disorder: Inventing a Disease to Sell Low Libido," *Journal of Medical Ethics* 41, no. 10 (2015): 859–62. For a more recent discussion, see Maxime Charest and Peggy J. Kleinplatz, "A Review of Recent Innovations in the Treatment of Low Sexual Desire," *Current Sexual Health Reports* 10, no. 4 (2018): 281–86. **This is not to deny that pharmacological:** Kristina Gupta, "Protecting Sexual Diversity: Rethinking the Use of Neurotechnological Interventions to Alter Sexuality," *AJOB Neuroscience* 3, no. 3 (2012): 24–28, 26. In the case of Addyi, one major question facing the FDA was whether the benefits outweighed the side effects of the drug. A substantial proportion of the women taking Addyi experienced a form of depression, dizziness, nausea, fatigue, somnolence, or sedation. There was also the risk of fainting or accidental injury, as well as potential adverse interactions with alcohol and common medications, including antidepressants (SSRIs) and hormonal contraceptives. Women certainly need to be informed of these risks. But in the end, it is each woman in consultation with her doctor who will have to decide if the risks outweigh the benefits in light of her own situation and values when she is paying for the drug. **would not themselves be a patentable source of profit:** See, however, the investigative reporting by Olivia Goldhill, "A Millionaire Couple Is Threatening to Create a Magic Mushroom Monopoly," *Quartz*, November 8, 2018, https://qz.com/1454785/a-millionaire-couple-is-threatening-to-create-a-magic-mushroom-monopoly/. **Current pharmacological approaches in psychiatry:** Ben Sessa, pers. comm., September 27, 2017. The other quotes from Sessa in this chapter are also from this interview. Quoted with permission. **Emerging in its place:** Matthis Synofzik, "Ethically Justified, Clinically Applicable Criteria for Physician Decision-Making in Psychopharmacological Enhancement," *Neuroethics* 2, no. 2 (2009): 89–102, 90; see also Lauren C. Heathcote, Daniel S. Goldberg, Christopher Eccleston, Sheri L. Spunt, Laura E. Simons, Louise Sharpe, and Brian D. Earp, "Advancing Shared Decision Making for Symptom Monitoring in People Living beyond Cancer," *Lancet Oncology* 19, no. 10 (2018): e556–e553. **In addition, the power of pharmaceutical companies:** For other examples, see Ben Goldacre, *Bad Pharma: How Drug Companies Mislead Doctors and Harm Patients* (New York: Random House, 2013). **Social movements, grassroots efforts:** Joseph E. Davis, "Medicalization, Social Control, and the Relief of Suffering," in *The New Blackwell Companion to Medical Sociology*, ed. William C. Cockerham (Singapore: Wiley-Blackwell, 2010), 211–41, 216. **The ethicist Kristina Gupta:** Gupta, "Protecting Sexual Diversity. . . ," 27. To these measures Rebecca Bamford would add the following: (1) regular retraining for healthcare providers and revisiting of the possible effects of unconscious biases; (2) monitoring of those licensed to supply the relevant technologies; (3) accountability to nongovernmental agencies involving members of relevant minority communities; (4) widening other aspects of diversity education and training to incorporate attention to economic, ethnic, religious, age, disability, and linguistic diversity; (5) increasing public understanding of the value of arts and humanities education and research to understanding the diversity of conceptions of love as these are relevant to health; and (6) increasing government funding for such research. Rebecca Bamford, "Unrequited: Neurochemical Enhancement of Love," *Cambridge Quarterly of Healthcare Ethics* 24, no. 3 (2015). **Recognizing this, ethicists are:** Synofzik, "Ethically Justified, Clinically Applicable Criteria . . . " **The old specter of pervasive medical surveillance:** This paragraph and some others in this chapter are adapted from Brian D. Earp, Anders Sandberg, and Julian Savulescu, "The Medicalization of Love," *Cambridge Quarterly of Healthcare Ethics* 24, no. 3 (2015): 323–36. **For example, by focusing completely:** Conrad, *Medicalization of Society*, 152. **Framing complex psychological phenomena:** The following text is adapted from Julian Savulescu and Brian D. Earp, "Neuroreductionism about Sex and Love," *Think* 13, no. 38 (2014): 7–12. **Here is a passage:** "Libido Problems: Brain Not Mind," *BBC News*, October 26, 2010, www.bbc.com/news/health-11620971. **For example, a brain tumor may cause an increase:** Charles Choi, "Brain Tumor Causes Uncontrollable Paedophilia," *New Scientist*, October 21, 2002, www.newscientist.com/article/dn2943-brain-tumour-causes-uncontrollable-paedophilia/#.U3EBgF%205H1fM. **So even if the cause is biological:** Synofzik, "Ethically Justified, Clinically Applicable Criteria . . . ," 95. **The mode or type of intervention:** Julian Savulescu, Anders Sandberg, and Guy Kahane, "Well-Being and Enhancement," in *Enhancing Human Capacities*, ed. Julian Savulescu, Ruud ter Meulen, and Guy Kahaner (Oxford: Wiley-Blackwell, 2011), 3–18. **"I feel so much better:** Jonah Lehrer, "Depression's Upside," *New York Times Sunday Magazine*, February 28, 2010, as cited in Parens, "On Good and Bad Forms of Medicalization." **As the philosophers Diana Aurenque:** Diana Aurenque and Christopher W. McDougall, "Amantes Sunt Amentes: Pathologizing Love and the Meaning of Suffering," *American Journal of Bioethics* 13, no. 11 (2013): 34–36, 35. **Similarly, the feminist bioethicist Laura Purdy:** Laura Purdy, "Medicalization, Medical Necessity, and Feminist Medicine," *Bioethics* 15, no. 3 (2001): 248–61, 256. **And when that is the case:** Purdy, "Medicalization . . . ," 255. **Combined with efforts to address the social factors:** Kristina Gupta, "Anti-love Biotechnologies: Integrating Considerations of the Social," *American Journal of Bioethics* 13, no.

11 (2013): 18–19, 19. **The point of the experiment was to see:** Arthur Aron, Edward Melinat, Elaine N. Aron, Robert Darrin Vallone, and Renee J. Bator, "The Experimental Generation of Interpersonal Closeness: A Procedure and Some Preliminary Findings," *Personality and Social Psychology Bulletin* 23, no. 4 (1997): 363–77. **They invited the entire lab:** Mandy Len Catron, "To Fall in Love with Anyone, Do This," *New York Times*, January 9, 2015, www.nytimes.com/2015/01/11/fashion/modern-love-to-fall-in-love-with-anyone-do-this.html. **Catron's article went viral:** For the TED talk, see https://tinyurl.com/y86rhy4s. **In his 1956 masterpiece:** Erich Fromm, *The Art of Loving* (New York: Harper Brothers, 1956). **As we quoted in the book's epigraph:** Fromm, *Art of Loving*, 56.

Epilogue: Pharmacopeia

Modern psychiatry is characterized by: Ben Sessa, *The Psychedelic Renaissance: Reassessing the Role of Psychedelic Drugs in 21st Century Psychiatry and Society* (Herndon, VA: Muswell Hill Press, 2012), 3. **He argues that the current "pharmacopeia":** Sessa, *Psychedelic Renaissance*. **"I care passionately about psychedelics:** Ben Sessa, pers. comm., September 27, 2017. Quoted with consent. **"Half the things I prescribe to people:** Quoted in Natasha Preskey, "Could MDMA Save Your Relationship?" *Elle*, July 21, 2017, www.elleuk.com/life-and-culture/culture/longform/a36937/could-mdma-save-your-relationship/.

INDEX

ABOUT THE AUTHORS

BRIAN D. EARP is Associate Director of the Yale-Hastings Program in Ethics and Health Policy at Yale University and the Hastings Center, and a Research Fellow in the Uehiro Centre for Practical Ethics at the University of Oxford.

JULIAN SAVULESCU holds the Uehiro Chair in Practical Ethics at the University of Oxford and directs the Oxford Centre for Neuroethics.